COMPLEXITY THEORY FOR A SUSTAINABLE FUTURE

COMPLEXITY IN ECOLOGICAL SYSTEMS

COMPLEXITY IN ECOLOGICAL SYSTEMS

TIMOTHY F. H. ALLEN and DAVID W. ROBERTS, editors
ROBERT V. O'NEILL, adviser

COMPLEXITY THEORY
FOR A
SUSTAINABLE FUTURE

Edited by Jon Norberg and Graeme S. Cumming

COLUMBIA UNIVERSITY PRESS   NEW YORK

COLUMBIA UNIVERSITY PRESS
Publishers Since 1893
New York Chichester, West Sussex

Library of Congress Cataloging-in-Publication Data

Complexity theory for a sustainable future / edited by Jon Norberg and Graeme S. Cumming.
　　p.　cm.—(Complexity in ecological systems)
　Includes bibliographical references and index.
　ISBN 978-0-231-13460-6 (cloth : alk. paper) — ISBN 978-0-231-13461-3 (pbk. : alk. paper)
— ISBN 978-0-231-50886-5 (e-book)
　1. Social ecology.　2. Sustainable living.　3. System theory.　4. Environmental policy.
I. Norberg, Jon.　II. Cumming, Graeme S.　III. Title.　IV. Series.
　HM861.C66 2008
　304.201—dc22

2007052041

Casebound editions of Columbia University Press books are printed on permanent
　and durable acid-free paper.

Printed in the United States of America

References to Internet Web Sites (URLs) were accurate at the time of writing. Neither the
　author nor Columbia University Press is responsible for Web sites that may have expired
　or changed since the book was prepared.

To all people who find joy in improving our common future

CONTENTS

FOREWORD

Simon Levin

HUMANS HAVE MANAGED parts of the natural world for millennia, harvesting trees and fish, extracting resources, and converting land into agriculture with an arrogance and naïveté that these represented linear perturbations of an infinitely forgiving planet that could continue to provide humans all they wanted, no matter how large their appetite grew. Late in the game, respect is building for the need to manage sustainably; this book is inspired by the conviction that it's not yet too late, but that time is short for addressing the problems.

Of course, the notion of sustainability is not new. In the theory of fisheries, for example, concepts such as maximum sustained yield and maximum sustained rent have been the centerpieces of a rich mathematical development, although they have been as often vilified as praised. From the beginning of the development of the theory of fisheries it was recognized that management practice was fundamentally challenged by the fact that we live in a global commons. Open access fundamentally undercuts prudent resource management and makes the simplest optimization goals impossible to achieve. But recognition that humans live in complex societies did not carry over to a similar recognition that fish and other resource species live in complex ecosystems or the corollary that management of these resources must be from an ecosystem perspective. The collapse of so many marine fisheries has now made ecosystem management an obvious imperative, but it should always have been so.

With this background, it has become increasingly clear that linear approaches, which treat each species as independent and each manipulation as occurring in isolation, are insufficient; nonlinear interactions within coupled natural and socioeconomic systems change the nature of the game.

In complex systems, prediction is much more problematic, and the interplay among diverse subsystems enhances the potential for dramatic regime shifts that fundamentally alter the nature of the larger systems. Fisheries collapse, epidemics erupt, lakes become eutrophied, and rich land becomes desert. The classical approaches in mathematical ecology, focusing on the linear stability of point equilibria, are insufficient to address these far-from-equilibrium phenomena. More relevant are issues of global stability and structural stability, which erect difficult challenges for quantitative approaches and for management.

Buzz Holling, in a deservedly influential work three decades ago, established the framework that is changing the way resource systems and ecosystems are managed. Holling emphasized the importance of the potential for sudden transitions of systems from one basin of attraction to another; that is, for shifts that fundamentally change the qualitative nature of systems. He turned attention away from stability to small perturbations and to the broader issue of the resilience of systems to large perturbations; that is, to the capacity of the system to continue functioning in qualitatively unchanged ways in the face of stressors. Holling's theoretical ideas led to the later development of the Resilience Alliance and also form a point of departure for this text.

Ecosystems, the biosphere, and the socioeconomic systems within which they are managed, are all particular kinds of complex systems. Indeed, like any complex systems, they are composed of components that interact nonlinearly, giving rise to the potential for periodic and chaotic dynamics, multiple stable basins of attraction, and hence regime shifts. But, beyond that, they share the feature that they are collectives, made up of individual (*microscopic*) agents that have their own selfish agendas and interact locally in space and time and that the outcomes of those interactions lead to dynamic changes in the relative frequencies of types. These changes can occur simply through the plasticity of animal behavior or physiology, including humans, or may represent genetic replacement through natural selection. Whichever is the case, the collective consequences are manifest in shifts in the *macroscopic* features of the systems in which they occur. Hence these are all examples of *complex adaptive systems*, whose properties change because of the interplay between the generalized adaptive responses of the parts and the emergent properties of the whole.

The study of complex systems has seen a flowering in the last several decades, and its historical development is well documented in this book.

Norberg and Cumming, however, have provided the first complete effort to build a bridge between the large body of theory that has emerged from places like the Santa Fe Institute and the application of that theory to achieving a sustainable future. The editors are to be commended for their seamless integration of resilience theory, sustainability concepts, and applications to specific case studies. The distinguished set of authors have provided state-of-art treatments of the translation and application of concepts from the theory of complex adaptive systems to coupled social-ecological systems, from the diversity and flows in ecological communities to the emergence of norms in human societies. This book is a unique contribution to the literature, which will serve as a pillar in the development of a new science of sustainability.

PREFACE

THIS BOOK HAS ITS ORIGINS in three influential lines of research: the Resilience Alliance program, initiated by C. S. ("Buzz") Holling and collaborators, Simon Levin's work on complex systems in ecology and resource management, and the Santa Fe Institute's program on complexity.

Buzz Holling founded the resilience network, later to become the Resilience Alliance (RA, www.resalliance.org). The RA is a multidisciplinary research group that explores the dynamics of complex adaptive systems (CAS). Its mode of operation is to facilitate cooperation among top researchers across disciplines using various arenas of communication. One way of achieving continuous evaluation, assessment, and reanalysis of the concepts discussed during RA scientific meetings was for the organization to invite and promote a group of young scientists (which also included the editors of this book). The goal was to create an opportunity for learning and continuity as well as to solicit constructive criticism from scientists who had not been exposed to the organization's history of thought or indoctrinated into it. The seed of the present book was planted by this group of young scientists at a meeting in which we tried to synthesize a set of ideas that had been further developed during the previous two years of active participation in the RA.

Buzz Holling's book on adaptive assessment and management in 1978 developed a number of the principles that today underlie the conceptual basis of adaptive management. Holling applied these principles not only to the natural systems that were the target of management efforts but also to the social process of management itself. His awareness of the importance of maintaining strategic options, his appreciation of continuous change, his recognition of the limitations of our understanding of nature, and his arguments in favor of the creation of feedbacks between management actions and their outcomes when implemented was far-reaching and continues to strongly influence policy making today. Most of the authors in this volume have also been influenced, to various degrees, by two other research agendas: some of the groundbreaking ideas coming out of the Santa Fe Institute (SFI) and the influential writings of Princeton professor Simon A. Levin. Early complexity theory did not necessarily focus on adaptation, but was a term used for nonlinear systems that produced complex dynamics from simple rules. People associated with SFI, such as John Holland and Stuart Kauffman, emphasized the idea of self-organization and the importance of adaptive processes for the development of complexity. Simon Levin's work has further stressed the importance of adaptation and its scale dependence both in the evolution of organisms and in a more general sense of adaptation as an important consequence of the action of selection on diversity.

These three strands of research come together in interesting ways throughout this volume. An important consequence of nonlinearity among system components is the potential for regime shifts, which describe a fundamental change in the underlying processes that govern system dynamics. The potential for regime shifts in social-ecological systems has vast implications for how these systems can and should be managed. To what extent are complex adaptive systems capable of dealing with change, and what determines their capacity to cope? Are there ways in which managers can buffer their system against regime shifts, keeping it in desired states and avoiding destructive change? These questions are the point of departure for ideas about system resilience, the capacity of a system to be subjected to disturbance without shifting into a new regime, and the focus of the Resilience Alliance.

This book takes three theoretical perspectives and one practical perspective on complex systems theory. Complex systems are first viewed as asymmetrical systems that dissipate energy and are composed of a variety of components. Second, they are viewed as networks of interacting nodes and links; third, as systems that process information in the sense of receiving external inputs and responding to them with outputs. Last, and somewhat uniquely in a book about complex systems, they are viewed from a management and policy perspective. Each of these perspectives provides a different way of organizing ideas about complexity that help us to conceptualize and visualize complex system problems. Part 4 is an attempt to spell out some of the implications of complex systems theory for managers and policy makers and to demonstrate how some of the more theoretical ideas discussed in earlier chapters can have important practical consequences for natural resource management. While the application of complex systems theory to problems of natural resource management is still in its infancy, this is an exciting and fast-moving frontier in science.

We hope this book can be seen as a complement to the inspirational book *Panarchy* published by the RA in 2001. It has inherited many of the ideas first proposed in *Panarchy*, while expanding on some of them and introducing others that have arisen from different perspectives. The diversity of perspectives from which CAS theory is developing is one of its strengths, because differences between conceptual frameworks and disciplinary backgrounds continually trigger communication between scientists and have been the source of many new ideas. We hope that the present book contributes to the further development of complex systems theory as an exciting interdisciplinary body of theory, while pointing toward the many applications of CAS theory to some of the most pressing problems of our time.

COMPLEXITY THEORY FOR A SUSTAINABLE FUTURE

INTRODUCTION

Jon Norberg and Graeme S. Cumming

THE STORY of how life on earth developed is about a gradual increase in the complexity, organization, and information-processing capabilities of organisms, from single cells with largely reactive behaviors to interactive multi-celled organisms with elaborate senses, learning capabilities, and proactive behavior. This story is paralleled in the development of human societies, which have grown from autonomous foraging groups of early humans to highly interconnected societies with specialized individual occupations and livelihoods, reduced local uncertainties due to trade and information exchange, and elaborate institutional arrangements for governance. The similarities between individual and social development are striking, and suggest the existence of a more general set of principles. Parallels between social and individual evolution must, however, be drawn with care; although Charles Darwin's remarkable formulation of the process of evolution by means of natural selection provided the understanding needed to explain the evolution of life on earth, naive analogies between natural evolution and social development have resulted in misguided conclusions and horrific results. Social Darwinism, for instance, has had a strong and negative influence on attempts to explore the conceptual links between evolutionary and sociological theories.

The coining of the term complex adaptive systems (CAS) by a group of scientists led by John H. Holland, Murray Gell-Mann, and others at the Santa Fe Institute (founded in 1984) provided the kind of value-neutral concept or metaphor that was needed to guide ideas and understanding about the common phenomena seen in many different areas of research. Using such a concept to look at complex phenomena in societies and natural systems has revealed that CAS thinking has emerged in many disciplines and that the processes that underlie CASs seem to be common to many different systems. For example, the process by which selection among options leads to self-organization and large-scale phenomena is found in such diverse fields as economics (Arthur et al. 1997), stock markets and manufacturing businesses, institutional arrangements (Lansing 2003), political sciences (Ostrom 1998, 2005), and ecosystems (Levin 1998, 1999; Holling 1973, 1992). This level of generality provides a strong indication that interdisciplinary commonalities exist, and that understanding such processes may be immensely useful for making better decisions in the future.

CASs are made up of interacting components (the system) whose interactions may be complex (in the sense of nonlinear) and whose components are diverse and/or have a capacity for learning that generates reactive or proactive adaptive behavior. Complexity theory preceded the idea of CASs; the CAS concept is set apart by its focus on the term *adaptive*, which, in addition to complex-only systems, explicitly focuses on the capacity of the system to change in response to prevailing (and anticipated, where possible) conditions by means of self-organization, learning, and reasoning. A change in a person's behavior following an experience can be termed adaptive if the same person, facing the identical situation as before, makes a different decision based on improved understanding. Similarly, an entire ecosystem can be said to adapt in the sense that individual system components can collectively respond to an external force. For example, the composition of the arctic vegetation is shifting toward a greater proportion of warmth-adapted species in response to climate change.

Complex systems are systems that can produce unexpected dynamics because of nonlinear interactions among components. Nonlinear dynamics mean that these systems have the possibility of self-reinforcing or self-moderating processes. One phenomenon caused by self-reinforcing processes is regime shifts, which refer to the ability of a system to internally switch between different self-reinforcing processes that dominate how the system functions. In contrast to the self-organizational process, this does not mean that the components of the system need to change. Rather, it conveys a change in the proportional contribution of different processes to the structure of the entire system. The possibility for regime shifts has fundamental implications for humanity's capacity to manage its life-support systems. Systems that tend to recover whenever one recognizes and removes a bad management practice are easy to manage by trial and error. However, systems that are fundamentally altered by a regime shift do not bounce back, and learning by trial and error becomes harder. The concept of resilience is used to describe the degree of disturbance a system can handle before a regime shift takes place. While the possibility for regime shifts as defined above is not necessarily present in all systems, whether natural or social, there is a special class of systems that exhibit multiple stable states and have proven very hard to manage. Furthermore, since the phenomenon of regime shifts was identified, the number of identified multiple regimes in social and ecological systems is increasing rapidly (Walker and Meyer 2004).

While regime shifts are potentially only a special kind of CAS dynamic, the idea of self-organization leading to adaptive processes applies to a wider range of systems. Importantly, adaptive processes relate to the system's capacity to tolerate change (i.e., response capacity), and hence to the ways in which systems switch between different regimes. System dynamics after a regime shift are also crucially dependent on the self-organizing capacity of the system. This is why the concept of resilience incorporates the ideas of adaptation, learning, and self-organization as a fundamental source of resilience in addition to the general ability to tolerate disturbance. Resilience can thus be described (Carpenter et al. 2001) as

1. the amount of disturbance a system can absorb while still remaining within the same state or domain of attraction
2. the degree to which the system is capable of self-organization (versus lack of organization or organization forced by external factors)
3. the degree to which the system can build and increase its capacity for learning and adaptation

More precise definitions have been attempted, but forcing a more narrow interpretation may decrease the usefulness of the concept. It is important to understand that resilience is a concept or perspective that is intended to guide thought rather than a scientific hypothesis that should be tested with quantitative data. This distinction does not mean that *aspects* of resilience cannot be measured, such as the value of a threshold and how far the system is from it or the ability of a species community to shift in response to climatic change.

Sustainability is another term that serves a useful purpose, but has, like resilience, many different aspects. The 1987 Brundtland Report defined sustainable development as development that "meets the needs of the present generation without compromising the ability of future generations to meet their needs." In this book we think of sustainability as the equitable, ethical, and efficient use of social and natural resources. Equitable here means equitable among people with different economic or social status, from different parts of the earth, or between today's and future generations. By ethical we mean in concordance with today's (and future) human values. By efficient, we mean that decisions about what to do should be based on the premise of providing the least resource-greedy solution constrained by the above criteria. We use this definition because it more explicitly stresses the importance of the underlying human aspects of sustainability; justice, human values, and efficiency in addition to the maintenance of long-term function in natural systems.

While resilience and sustainability (and associated research programs) provide useful ways to conceptualise and grapple with current problems, it seems clear that to understand resilience and sustainability in social-ecological systems one needs to appreciate that they are complex adaptive systems. One of the most useful characterizations of CASs comes from S. Levin's paper (1998) and later book *Fragile Dominion* (Levin 1999) in which he lists the following as core ingredients of CASs: (1) sustained diversity and individuality of components; (2) localized interactions; and (3) an autonomous selective process.

In order to generalize these themes for more transdisciplinary use, with special application to sustainability research, we choose to re-word them in the terms of established theoretical bodies of research. We therefore identify the following ingredients in social-ecological systems that make them good representatives of complex adaptive systems.

DIVERSITY AND ASYMMETRY   The world is essentially uneven, with variable environments in space and time that drive important processes. Even though patterns can theoretically emerge in featureless environments (symmetry breaking), large asymmetries in climatic drivers, economic wealth, or information sustain processes such as speciation, trade of services, or collective action. Thus variability can be the cause of processes that sustain a diversity of system components. Systems also adapt to strong selective processes, which act on diversity; diversity itself therefore plays a large role in how systems perform. In this volume we present one chapter (1) on the role of spatiotemporal variability in environmental drivers as a cause of processes that structure social-ecological systems and one chapter (2) that addresses the role of diversity in the performance and resilience of social and ecological systems. The interaction between these two aspects is addressed in a chapter (9) on scale-related issues and cross-scale interactions. Chapter 4 elaborates further on the themes of diversity and asymmetry by discussing the importance of diverse actor groups in social networks for creating change and for the transformation of old paradigms of management. On a similar note, chapter 7 brings in the importance of stakeholder participation in formulating scenarios of regional development.

NETWORKS   Localized interactions are the basis of the general concept of networks. Unless all nodes in a network interact equally with all other nodes, some degree of localization takes place. The degree and quality of this localization is captured by the topology of a network, i.e., the strength and dynamics of the links that connect the nodes. Furthermore, the quality and behavior of different nodes fundamentally determines system performance. Network theory is an established concept in both social and biogeophysical sciences and provides a substantial amount of common ground between different disciplines. Chapter 3 offers a deep yet general review of network theory as it applies to sustainability issues, while chapter 4 addresses the role of linking existing networks in natural resource management under a common vision to achieve a transformation of thought.

INFORMATION PROCESSING   Information processing is a complex topic; its relevance as a section title needs some additional explanation. Selective processes are a result of differential responses of components to some pressure. If these responses result in increased relative abundance or use of a component, then this type of component is selected for. This is the fundament of natural selection; the responses of species to environmental drivers lead to evolution and adaptation within communities of species. In social systems the process of choosing between alternative options involves processing of information about the expected outcomes of different options and determining how to proceed - i.e., making decisions. Whether conscious or unconscious, constitutional, collective, or individual, decisions are the main selective process in the social system. But information processing is also a general term used in physics and relates to how a system, or system components, responds to an input signal. Such a response could be, for example, the varying responses of arctic versus temperate plants to temperature, which lead to a directional change in the characteristics of arctic vegetation toward those of temperate vegetation as a result of climatic change.

Another such nonlinear response is the rate of phosphate regeneration in lake sediments as a response to oxygen levels. Nonlinear responses may cause complex dynamics, as discussed earlier. In each of these instances, the system is effectively processing information and responding through modification of its properties or behaviors. Information processing therefore offers a conceptual handle that captures a diverse range of responses, emphasizing commonalities between them.

In the section on information processing, Chapter 5 addresses how information about the world is processed into guided actions and the fundamental problems of having to rely on a simplified representation of the real world when making decisions. Chapter 6 provides compelling evidence that suggests that the variance in important system variables increases as a system approaches a transition (state change, or threshold), offering a potentially valuable analytical tool for managers who are monitoring systems that are approaching a potential regime shift.

In assembling this volume, we have tried to use value-neutral terminology that is transferable between the social and biogeophysical sciences in order to establish a better ground for communication. This does involve some compromises and requires a willingness to see one's own field from the perspective of other fields of research. Our main aims with this book are to contribute to the science of sustainability, to provide a framework for better understanding of our life-support systems, and ultimately to facilitate the development of better decision making and more effective policies. We think that complexity theory and complex adaptive systems theory has been sufficiently developed (and has gained enough acceptance within the scientific community) for it to be put to use. This book is therefore intended to first explore the main aspects of CASs from a sustainability perspective, and then to complement this theoretical exploration with a collection of chapters that particularly address practical approaches and the applications of insights gained from the foundation of CAS thinking. Lastly, we hope that this volume will help to bridge some of the gaps between different disciplines and to stimulate the development of new questions and ideas.

REFERENCES

Arthur, B. A., Steven N. Durlauf, and David A. Lane, eds. 1997. *The Economy as an Evolving Complex System II.* Vol. 27. Santa Fe Institute Studies in the Science of Complexity. Reading, MA: Addison-Wesley.

Carpenter, S., B. Walker, J. M. Anderies, and N. Abel. 2001. "From Metaphor to Measurement: Resilience of What to What?" *Ecosystems* 4:765–781.

Holling, C. S. 1992. "Cross-scale Morphology, Geometry, and Dynamics of Ecosystems." *Ecological Monographs* 62(4): 447–502.

Holling, C. S., ed. 1978. *Adaptive Environmental Assessment and Management.* London: Wiley.

Lansing, J. S. 2003. "Complex Adaptive Systems." *Annual Review of Anthropology* 32:183–204.

Levin, S. A. 1998. "Ecosystems and the Biosphere as Complex Adaptive Systems." *Ecosystems* 1:431–436.

Levin, S. A. 1999. *Fragile Dominion.* Reading, MA: Perseus.

Ostrom, E. 1999. "Coping with Tragedies of the Commons." *Annual Review of Political Science* 2:493–535.

Ostrom, E. 2005. *Understanding Institutional Diversity.* Princeton: Princeton University Press.

Walker, B., and J. A. Meyers. 2004. "Thresholds in Ecological and Social-Ecological Systems: A Developing Database." *Ecology and Society* 9:3. http://www.ecology andsociety.org/vol9/iss2/art3.

INTRODUCTION TO PART 1
DIVERSITY AND HETEROGENEITY

IN THIS FIRST SECTION, we focus on the role of diversity in complex adaptive systems. Diversity is called by many names in science; we interpret its meaning broadly here, equating it with differentiation, heterogeneity, and variation. The central interrelated themes that are explored in this section are 1. the role of diversity in sustaining processes that drive and maintain the system; and 2. the processes that sustain and diminish diversity itself.

Diversity is one of the most obvious patterns that arises in complex systems. The study of diversity has had an enormous influence on biology, particularly in ecology and evolution. Ecological definitions of species diversity invoke two main facets of an ecological community, the number of species present and the number of individuals of each species. More diverse communities have more species and a more even distribution of individuals among species. Diversity is also an important concept in the human sciences, for example in studies of equity and wealth distribution, investment strategies, population demographics, languages, or institutional diversity. An important aspect of diversity is the nature of the differences between system components and the ways that they relate to one another. Each individual component of a complex system may have many properties and many functions; for example, an antelope is simultaneously a grazer, a food source, a disperser of nutrients, and a producer of methane. Any component of a complex system may have some properties that are redundant (i.e., that are duplicated by other system components) and others that add diversity to the system.

It is no accident that the concept of diversity is useful in a wide variety of instances. One of the few generalizations that we can reliably make about complex adaptive systems is that complex *adaptive* behaviors depend on some level of differentiation of system components. Complex, nonlinear behavior is possible without component diversity, for instance in a turbulent but otherwise homogeneous fluid, but adaptation relies on variations and differences between system components. Diversity means more than simply having a range of different individuals, strategies, or populations. From the perspective of an entire system, diversity means having a greater range of *options* for responding to environmental change and a correspondingly higher likelihood that a solution to a particular problem will be found.

When we try to understand the role of diversity in complex systems, we are particularly interested in three kinds of question: 1. Where does diversity come from? How does it arise, and what maintains it in the system? 2. What reduces or removes diversity from the system? Under what circumstances is homogeneity favored? And 3., what is the functional role of diversity? What kinds of diversity make the largest contribution, positive or negative, to systemwide properties of interest, such as persistence or resilience? We elaborate a little on each of these questions and then introduce the two chapters that follow.

ORIGIN AND MAINTENANCE OF DIVERSITY

The origin of novelty in complex systems has long been one of the least understood aspects of diversity. Charles Darwin's mentor, the geologist Charles Lyell, once compared the theory of evolution to the supreme Hindu triad of gods; the action of natural selection as Shiva the destroyer and Vishnu the sustainer is self-evident, but he found it difficult to see how selection could also act as Brahma, the creator. Diversity in complex adaptive systems arises by chance and imperfection, recombination, and in social systems, by innovation and foresight. In DNA, for example, random mutations are often caused by errors in decoding and copying the genetic code. New behaviors, such as a pack of lions working out that giraffes are more likely to fall if chased along the smooth surface of a road, are often discovered by chance. The teabag, an invention that was to prove incredibly successful, was introduced by mistake when rich customers failed to realise that they were expected to cut open the silk envelopes in which their tea was purchased. New institutions may be invented by transformation of knowledge from unrelated domains, such as the introduction of tradable quotas in fisheries. The origin of novelty is not independent of its environment; rapid change may be more likely under certain circumstances, such as when environmental conditions favor physical processes that foster novelty or force societies to develop and test new solutions that might otherwise have been considered too expensive.

The processes and conditions that promote diversity are discussed in chapter 2. They include spatiotemporal changes that prevent the domination of complex systems by a single entity; resource aspect differentiation, as, for example, diversifying allowable types of seafood harvested to decrease sensitivity to population dynamics in single species; and deliberate management of key components that sustain diversity in the system, such as keystone species that feed on the dominant competitor, or enabling legislation that directs collective action to local users.

Once novelty has arisen, it can travel rapidly through a system. Richard Dawkins has captured this concept elegantly with his descriptions of genes and memes, both of which constitute replicating units that multiply faster when they are well suited to their environment. Diversity is maintained by successful innovations that replace those that are lost from the system through selection. It often increases in situations in which there is decreased competition between system components or reductions in other forms of selective pressure.

LOSS OF DIVERSITY

Diversity is reduced by selective processes, which are fundamental to CASs (Levin 1999). Selection favors one kind of system component over another, often based on a single critical variable that influences survival. For example, global climate change looks likely to favor coral polyps that can cope with warm water over those that require colder water. Selection is by no means restricted to the domain of ecology; companies and their products experience selective pressure from consumers, the value of art is closely related to critical opinion, and in the empirical sciences (with a few exceptions) experimental evidence becomes a selective criterion that is used to separate fact from speculation.

Selective processes are closely related to ideas about optimality. In a constant environment directional selection should drive adapting systems toward an optimal solution and then stabilize their location in evolutionary space. Stuart Kauffman and Simon Levin, among others, have visualized this theoretical space as a landscape in which systems move up a gradient of increasing efficiency or survivorship toward local optima, which may or may not also be global optima. This landscape is continually changing as the system and its environment fluctuate.

The search for efficiency in modern societies often translates into a desire to optimize. Under conditions when optimization is at a premium (i.e., when the future appears to be predictable), the maintenance of diversity seems unnecessarily costly. For example, as the world undergoes a transition from local markets and local production toward large multinational companies, efficiency is increased and economies of scale push prices down. At the same time, however, choice and diversity are being eroded. Policies are also being adapted to cover larger regions, a striking example being the transition from national governance to common European-wide institutions. These changes push companies and societies toward local optima, but may also make them increasingly vulnerable to environmental fluctuations as the scale of response increases. For example, homogenization and regional mixing of cattle herds, together with centralized abattoirs and a national meat distribution system, meant that the BSE (Bovine Spongiform Encephalitis) epidemic in the UK was far more difficult to control than it might otherwise have been.

CONSEQUENCES OF DIVERSITY

As the preceding discussion implies, diversity has a wide range of important consequences for complex adaptive systems. It can be equated with maintaining a range of options or set of answers that can be used to respond to questions that are posed by the environment. These options, together with the possibility to change between them, provide the adaptive capacity that characterizes CASs (Norberg et al., chapter 2, this volume). Thus diversity is most valuable during times of change, when the integrity of a complex adaptive system is threatened. During times of stability, diversity can become a burden on the system because maintenance of system components requires energy, and many components in a diverse system may be relatively inefficient or even unnecessary. Consequently, many complex adaptive systems go through cycles in which diversity is gradually reduced, which also reduces the adaptive capacity of the system and makes it more vulnerable to change. Disturbances may cause a loss of function and a systemic reorganization, possibly leading to a new configuration of the system.

A fundamental concept discussed in chapter 1 is the idea that environmental asymmetry or systematic variation in the environment is responsible for driving processes and 'making things happen.' Different kinds of asymmetry may interact in ways that produce complex system dynamics. For example, the long-range migrations of many species occur as populations take advantage of variations in resource availability. By moving, animals can adopt a strategy in which they both track resources and manage to avoid harsh conditions. Movement carries a penalty of increased risk and vulnerability, however, so other species—or sometimes, other individuals of the same species—adopt a different strategy and stay in the same area throughout the year. In this example, it is both the characteristics of the individual and the nature of the differences between variations in the abundance of predators and variations in the abundance of food that determine which solution will be most successful.

Asymmetry can also have important effects on human societies and our interactions with natural resources. For example, farm sizes in the United States have historically been more equitable than those in Latin American countries such as Brazil. These distinctions have consequences for land cover and land use; big and small farms are managed differently. Changes in management have consequences for animals and plants, and ultimately for the long-term sustainability of the system. Systematic variation in the human environment is thus important for many of the CASs that we are most interested in. The first chapter explores some of these ideas and concludes with a discussion of the possibility that causal interactions may occur between spatial and temporal asymmetries in the human-dominated landscape of north central Florida.

1

ENVIRONMENTAL ASYMMETRIES

Graeme S. Cumming, Grenville Barnes, Jane Southworth

THE WORLD IS essentially uneven. We exist in an environment that continually presents us with change and variation in multiple manifestations; sunlight and shadow, day and night, hills and valleys, town and country. Such systematic unevenness is an important property of complex systems, from the large to the small. In this chapter we explore the idea of environmental symmetry and its converse, environmental asymmetry. We start by defining what we mean by symmetry and demonstrating how symmetry concepts in a variety of guises have been widely applied in the biological and social sciences. We then consider the mechanisms that produce environmental asymmetry, the consequences of asymmetry, and the potential for interactions between different kinds of asymmetries in landscapes. We argue throughout that asymmetries are integral to self-organization in complex systems and are consequently of high importance for understanding complex system dynamics.

DEFINING ENVIRONMENTAL SYMMETRY

Although it is widely applied in physics, the current scientific usage of symmetry is unfamiliar to many people in other disciplines. We use symmetry in its modern group-theoretic sense to mean invariance of group members under a set of specified rotations or reflections (Brading and Castellani 2003). This definition includes the traditional concept of symmetry that most of us were taught at school; for example, a highly symmetrical human face is invariant under reflection in the vertical plane and an unblemished daisy can be considered invariant under rotation on the axis of its stem. However, the group-theoretic definition also extends the traditional concept of symmetry by including other kinds of transformation. Symmetry in this broader context is associated with equality of the parts with respect to the whole in the sense of an interchangeability of parts; it implies 'a unity of different and equal elements' (Brading and Castellani 2003).

Symmetries can be exact, approximate, or broken (Castellani 2003). Exact symmetries are unconditionally valid; approximate symmetries are only valid under certain conditions; and broken symmetry in our context implies a deviation from some theoretically plausible or expected symmetrical state. Symmetry breaking does not imply that no symmetry remains in the system, but rather that the state of the system is characterized by a lower symmetry than would be present where symmetry has not been broken (Castellani 2003). We use the term *asymmetry* to describe situations in which symmetry has been broken, making no distinction between asymmetry and alternative terms such as *nonsymmetry* and *dissymmetry*. In group-theoretic terms, symmetry breaking means that the initial symmetry group is broken into one (or several) of its subgroups. Asymmetries thus arise from symmetries rather than the other way around. In a system that contained absolute symmetry, nothing would exist; absolute symmetry means a complete lack of differentiation (Castellani 2003). Simon (1962) distinguished between hierarchical systems that are 'decomposable' (i.e., into discrete levels and subunits) and those that are not; highly symmetrical systems are generally more decomposable than highly asymmetrical systems and hence tend to be easier to model and to understand analytically. Nonetheless, many real-world hierarchical systems in which symmetry has been broken fall into Simon's 'nearly decomposable' category, implying that reasonable mathematical approximations of system dynamics can be achieved.

Since the beginnings of modern science, the homogeneity and isotropy of physical space and time have been taken for granted in most areas of research (Brading and Castellani 2003). A representation of space that uses a grid of coordinates or cells without attaching attributes to any location is symmetrical; a cell in any one location could be interchanged with any other. In a typical neutral landscape model in which cells are assigned a value of 0 or 1 using a random number generator, the resulting landscape is symmetrical in the sense that the overall properties of the landscape are invariant to spatial transforms. A landscape consisting of patches of three different types can be considered approximately symmetrical if random interchanges of patches between locations have little or no effect on the overall properties of interest within the landscape. By contrast, a landscape in which patches of the same three types are arranged along a gradient is asymmetrical. Note that the patches themselves are, by definition, expected to be symmetrical at a finer scale of analysis; when defining a patch, we assume that areas within the patch are interchangeable.

Symmetry questions come to the fore in the consideration of scale. Under a scaling transformation, symmetrical systems will be those that are fully or approximately scale invariant. For example, the relationship between log body mass and log metabolic rate in mammals is largely symmetrical; it can be described using a straight line (Peters 1983). Spatial analyses also use scaling relationships to assess how system properties change with changes in dimension. For example, the fractal dimension of a stream network can be described as the scaling function of the slope of the relationship between the number of cells containing part of the network and the dimension of the cells (Tarboton, Bras, and Rodriguez-Iturbe 1988). A straight line with a constant slope (and hence an isotropic fractal dimension) indicates symmetry in scale. Unsurprisingly, most real-world patterns are not self-similar across multiple scales, and hence can be termed asymmetrical. Scale in complex systems is discussed further in chapter 9.

Heterogeneity refers to the differentiation of landscape elements (at any given scale) in space. Variation, in its commonest usage, is effectively heterogeneity in time. Asymmetry has much in common with heterogeneity but is not identical to it. Heterogeneity implies differentiation of system components but not necessarily asymmetry. Asymmetry refers only to systematic differentiation—so in thinking about landscape asymmetry we focus on the role of nonrandom spatial and temporal differentiation (heterogeneity, variation) in ecological systems. The importance of differentiation of individual system components is discussed in more detail in the next chapter. Classical examples of landscape asymmetry include such things as the systematic arrangement of different vegetation types down an elevation gradient (spatial asymmetry), successional change in a mosaic of abandoned fields (temporal asymmetry), and continental rainfall patterns over a decade (spatial and temporal asymmetry).

Although we may appear to be proposing a novel application of the term *asymmetry* in relation to social-ecological systems, we are simply recognizing previous research on asymmetries and recasting it in a slightly more general framework. As we now demonstrate, symmetry concepts have a long history of application in many disciplines, including ecology, medicine, and economics.

APPLICATIONS OF SYMMETRY CONCEPTS IN THE STUDY OF COMPLEX SYSTEMS

Assumptions about symmetry and symmetry breaking are widely applied outside of physics, even though the philosophical basis for these assumptions is not always made transparent. We have grouped the applications of symmetry-related concepts into three different categories.

ASYMMETRY AS A CONSEQUENCE OR A RESPONSE

Asymmetry is often seen as a consequence of nonrandom variation or a form of environmental response to process variation. In this context it is used consistently with the definitions given earlier in this chapter; an implicit assumption that symmetry is to be expected means that asymmetries demand explanation. For example, historical variations in the formation of oceanic crusts and the movements of tectonic plates produce asymmetries in bathymetry and elevation (Borisova and Kazmin 1993; Pushcharovsky and Neprochnov 2003; Toomey et al. 2002). Historical changes in the local environment and historical measurement inaccuracies can produce asymmetries in buildings and consequently are of interest to architects (Cummings, Jones, and Watson 2002; Lemoine 1986; Middleton 1989; Wilson Jones 2001). In studies of plant communities, attempts have been made to explain asymmetry in vegetative structures that include leaf canopies and branch structures ('crown asymmetry') as well as root systems (Logli and Joffre 2001; Rajaniemi 2003; Schwinning and Weiner 1998; vanderMeer and Bongers 1996). Asymmetries in the paired wing and tail feathers of birds and the organs or body sizes of other animals are used as indicators of quality or malfunction in studies of sexual selection and nutrition (Ahtiainen et al. 2003; Buyanovsky 1996; Van Dongen, Lens, and Molenberghs 1999; Vollestad, Hindar, and Moller 1999). Asymmetries in foraging patterns may be a response to environmental heterogeneity or a consequence of self-organization (Portha, Deneubourg, and Detrain 2002). In businesses asymmetries in returns may result from differences in past activity and the ensuing difference in activity costs (Acemoglu and Scott 1997). And in studies of the human brain, asymmetry in neuron activity has been suggested as a diagnostic feature of schizophrenia (Gruzelier et al. 1999). Asymmetry is referred to less explicitly in scaling studies, but deviations from expected scaling relationships typically demand explanation. For example, in the highly symmetrical relationship between the hoof area and the body mass of ungulates, the larger than expected hoof size of the camel is explained as an adaptation to walking on sand (Cumming and Cumming 2003) and in studies of urban boundaries, changes in fractal dimension have been proposed as indicators of societal mechanisms (Brown and Witschey 2003; Chen and Zhou 2003; White and Engelen 1993; Yizhaq, Portnov, and Meron 2004). Asymmetry may also be seen as an integral part of scale-independent processes, as described in studies of fractal branching patterns in arteries (Schreiner et al. 1997; van Beek 1997).

SYMMETRY AND ASYMMETRY AS EMERGENT PROPERTIES OF COMPLEX SYSTEMS

Both asymmetry and symmetry have been viewed as emergent properties of complex systems. In bee and ant colonies, asymmetries arise in the division of labor; symmetry breaking occurs when the numbers of foragers visiting two equally profitable food sources diverge (de Vries and Biesmeijer 2002; Portha, Deneubourg, and Detrain 2002). In mathematical models of the process of evolution, accumulated random changes ('noise') may generate system bifurcations and, hence, sympatric speciation (Stewart 2003). A similar symmetry-breaking response is visible in the difference in oil demand between developed and less developed countries (Dargay and Gately 1995).

ASYMMETRY AS A CAUSE OR DRIVER OF PATTERN

Asymmetry has been portrayed as a driver of pattern in a number of different contexts. In business the role of asymmetries of information and other resources in determining business success has received considerable attention (Amir and Wooders 1998; Cooper, Downs, and Patterson 2000; Miller 2003). For example, asymmetries in brand switching may explain aspects of purchasing behavior (Wedel et al. 1995). Similarly, asymmetries in the colony sizes of ants can be key drivers of the outcome of competition (Palmer 2003). While the generation of asymmetries by landscape processes has been relatively well explored (e.g., Turner, Gardner, and O'Neill 2001), the converse (i.e., the feedback from pattern to process) is considered less frequently.

As this categorization shows, the relevance of asymmetry in complex systems can vary; asymmetry may be an integral component of a process, an emergent property that arises through the action of one or more processes, or an inevitable (and possibly trivial) response to a process. In each of these cases, however, asymmetry provides us with useful information about the relationship between pattern and process in the system of interest.

THE RELEVANCE OF ENVIRONMENTAL ASYMMETRY FOR STUDIES OF COMPLEX SYSTEMS

Landscapes at any scale are composed of numerous locations in space. These locations can usually be considered as a distinct group or set, even if the sole commonality that they share is their proximity to one another. Consideration of landscapes as symmetrical or asymmetrical groups leads to an interesting perspective on the relationships between structure and function. Pierre Curie, working on symmetry and symmetry breaking in crystals in the year 1894, initiated the empirical study of asymmetry by asking which phenomena can occur in a physical medium that has specified symmetry properties. What is the importance of the relationship between the symmetry of a physical medium and the symmetry of the phenomenon occuring within it? Investigating this question led Curie to the conclusion that if a phenomenon is to occur in a medium, the symmetry of the medium must be lowered to the symmetry of the phenomenon by the action of some cause. Curie's conclusion implies strongly that symmetry breaking, or the formation of asymmetry, is what "creates the phenomenon" (Castellani 2003).

Our central argument through the rest of this chapter is that Curie's conclusion has a broad and general application in complex systems; environmental asymmetry is essential for the maintenance of the majority of ecological, sociological, and economic processes that are of interest in understanding sustainability. To demonstrate the importance of asymmetry in complex systems, we discuss a number of specific, detailed examples. These examples are intended to explore the proximate causes of environmental asymmetry, some of its consequences, and the ways in which different kinds of asymmetries (spatial, temporal, and scale related) may interact to affect system dynamics.

CAUSES OF ENVIRONMENTAL ASYMMETRY

Environmental asymmetry arises from processes that break environmental symmetry. If the degree of environmental symmetry is a description of pattern, symmetry is altered by the action of process. Phrased differently, we can say that systematic variation in the environment arises from three main sources: 1. processes that occur at different rates, frequencies or magnitudes in different locations (e.g., deposition of sand on coastlines), 2. processes that are constrained by existing environmental variation (e.g., fire will only burn where the fuel load is sufficient), and 3. processes of equivalent rates and magnitudes that occur out of spatial or temporal synchrony with one another (e.g., succession in a patch mosaic). Note that this summary is similar to, but slightly different from, Levin's (1976) classification of the causes of spatial heterogeneity in ecosystems as local uniqueness, phase differences, and dispersal. We assume that in the absence of differential process action, environments would remain symmetrical; environmental asymmetry demands explanation, environmental symmetry does not.

Broadly speaking, the processes that break symmetry in the environment may come from one of three causes: abiotic drivers (including geology, climate, or other purely physical processes), biotic nonhuman drivers (including interactions between organisms or between organisms and the abiotic environment), or anthropogenic drivers (including humans and the social and economic systems that they create). Asymmetries can also be explained as consequences of disturbances (such as fires, floods, and hurricanes) and historical variation in symmetry-breaking processes.

A considerable amount has been written on the causes of asymmetry in landscapes (reviewed in Turner, Gardner, and O'Neill 2001). Indeed, this has been one of the central themes of ecology. There have been innumerable studies of the relationship between the abiotic environment and the biotic environment, including the degree to which ecological communities are determined by their abiotic environment, the role of biotic processes in modifying or moderating the abiotic environment, and the variables that maintain spatial patterns in the environment. Consideration of spatiotemporal scaling relationships has long been considered one of the more important challenges facing ecology (Levin 1999). The relationships between anthropogenic, biotic, and abiotic variables have also been widely studied. The study of primarily anthropogenic causes of asymmetries in landscapes (such as political ecology and tenure systems) has been less extensive, but is a rapidly growing field.

As these comments should make clear, a thorough summary of the causes of asymmetry in landscapes would require book-length treatment. In this chapter we are more interested in the consequences of environmental asymmetry, to which we now turn, and the interactions and feedbacks between different kinds of environmental asymmetries.

CONSEQUENCES OF ENVIRONMENTAL ASYMMETRY

Environmental asymmetries are central to a large number of ecological processes. They may cause changes in community structure and create gradients and edges that drive the movements of water, minerals, and organisms through landscapes. By generating potential differences between areas they effectively create flows and act as channels that regulate flows once they are in motion. The influence of environmental asymmetries on movement is obvious in the case of a weary hiker traversing a mountainous landscape, trying to select the safest route that offers little resistance to her movements. Just as she skirts the rocky peaks of hills and follows the river in the valley until a suitable crossing point is reached, so the edges created by systematic variation in the environment can constrain and direct movement patterns of a multitude of species.

One of the most important environmental asymmetries in the abiotic environment is the change in available energy with changing latitude from the equator to the poles. This gradient is associated with systematic variation in rainfall and temperature and has been implicated as a major driver of patterns of biodiversity in ecosystems (Allen, Brown, and Gillooly 2002; Brown 1995). At a finer scale the elevation gradient that ranges from hills to lowlands has a profound structuring effect on plant and animal communities through its influences on local climate, substrate, and the availability of water. The communities of many organisms, from trees to rodents and birds, are stratified according to the elevation at which they occur (Lomolino 2001; Whittaker 1960). Streams and rivers in temperate environments show predictable transitions of substrates and biota from highlands to lowlands (Vannote et al. 1980); coastal vegetation composition and structure are largely determined by the environmental gradients imposed by dune formation (or erosion) and the influence of winds and mists from the sea; and, in the Sahel, distinct banding patterns in stands of shrubs and trees (the so-called tiger bush) arise in arid areas as a consequence of asymmetries in subsurface water flows (HilleRisLambers et al. 2001; Rietkirk et al. 2000).

Asymmetries in the social environment are also important. As societies have shifted from a basis of agriculture to industry to information, there has been a growing recognition of the importance of information as a resource (Toffler 1980). Information is the basic ingredient in decision making related to planning, conservation, development, and a host of daily decisions made by individuals. In developing countries, characterized by large asymmetries in wealth distribution, there are corresponding information asymmetries. For example, people with wealth and power have access to information on land markets, prices, and land regulations, while the poor do not have the resources to obtain this information. Solutions to the problem of inequity in some nations have given rise to the advancement of transparent and accessible property systems where all sectors of society have equal access. In this sense, information may be regarded as a public good (Deininger 2003) that promotes greater economic and social symmetry among rich and poor.

Environmental asymmetries occur in both space and time. In variable environments, such as African savannas, the spatial distribution of resources varies in a predictable manner through the course of a single year. The biomass of large herbivores in a given area is strongly correlated with rainfall (Coe, Cumming, and Phillipson 1976). Du Toit (1995) states that 'the primary ecological determinants of large mammal communities in African savannas are rainfall and soil nutrients, since these determine the quantity and quality of food available to large herbivores.' Among the notable consequences of asymmetrical seasonal and spatial variation in resources are the broad-scale migrations of a wide range of animals. Migratory species in Africa include birds, ungulates, carnivores, insects, fishes, and fruit bats (e.g., Berger 2004; Thirgood et al. 2004; Thomas 1983; Trinkel et al. 2004; Walther, Wisz, and Rahbek 2004; Ward et al. 2003). Many human societies, particularly those in arid environments, have also developed behavioral patterns (such as migration and food storage) that help them to cope with spatiotemporal asymmetries in the environment.

In each instance people and animals follow a gradient of resource availability that coincides with the timing of rainfall and related vegetation patterns. Asymmetries in vegetation and rainfall patterns in the Sahel can result in outbreaks of migratory locusts (Despland, Rosenberg, and Simpson 2004); in the Serengeti and Kalahari systems wildebeeste and other ungulates travel long distances each year to find water and grazing (Fryxell, Wilmshurst, and Sinclair 2004); and spatiotemporal variations in forest fruit abundance are thought to drive the annual migration of the straw-colored fruit bat *Eidolon helvum* from the Congo forests south into Zambia (Richter and Cumming 2006). Many birds migrate between Africa and Europe, including wading birds, raptors, and songbirds (Meyburg, Paillat, and Meyburg 2003; Walther, Wisz, and Rahbek 2004). In other parts of the world the seasonal migrations of wading birds, caribou, fruit bats, and monarch butterflies are all driven largely by environmental asymmetries, supporting our contention that symmetry breaking is an important driver of processes in terrestrial landscapes. The same principle is applicable in the oceans; whales, manatees, and turtles undertake seasonal migrations along resource and temperature gradients, and many fish species respond to temporary 'loopholes' of high resource availability and/or low predator abundance in the ocean, using these ephemeral resource asymmetries to boost recruitment in a given year (Bakun and Broad 2003; Best and Schell 1996). Migrant workers and refugees respond to their social-ecological environment in much the same way, finding secure or resource-rich pockets and escaping bad conditions through movement.

In addition to regulating communities and driving movements of animals, environmental asymmetries can cause particular areas to become net exporters or importers of organisms, water, or other substances such as minerals and wastes. In many cases ecological and human communities are maintained by subsidies or flows of substances that have their origin in other areas (Polis, Power, and Huxel 2004). Source and sink areas, defined as net importers or exporters of organisms respectively, are largely determined by asymmetries in environmental quality. In other instances environmental asymmetries lead to a continual flow of resources from one area to another. A classic example is the dependence of many large cities on processes that occur in the upper catchments of their water sources. For instance, the town of Melbourne in Australia obtains its drinking water from the Murray-Darling basin. Water salinity in many catchment areas is increasing as a consequence of salty groundwater and the rise in the water table that has resulted from clearing of native vegetation for sheep production (Keating et al. 2002; MDBC 1999). The state government has been forced to impose a salt credit system on smaller municipalities upstream, where the farmers in each subcatchment can only export a certain amount of salt and must modify their agricultural activities accordingly.

Depending on their nature and their context, environmental asymmetries will play a central role in creating edges and channels within landscapes. Edges occur where two distinct habitat types meet; for example, at the interfaces between forest and grassland, water and land, road and grassland, or rock and soil. The proportion of edge habitat in a landscape may be increased by anthropogenic activities (Cadenasso et al. 2003). Edges will often occur where strong environmental asymmetries exist, because edges are created by processes that show the kind of differential localization or concentration that is anticipated in an asymmetrical landscape. For example, the actions of weathering and subsurface processes on asymmetries in rock formations create cliff edges; water concentrates in the lower parts of landscapes to create the land-water edge; and fires burn only in areas with adequate fuel loads, creating patches in different successional stages. Once created, edges may serve as channels or corridors for the movements of organisms and their propagules (e.g., Machtans, Villard, and Hannon 1996).

INTERACTIONS AND FEEDBACKS BETWEEN ASYMMETRIES

In most environments multiple processes act to create multiple asymmetries. Different asymmetries will not necessarily follow the same gradient in space and time. For example, soil properties are influenced by weathering processes, and the climate that produces weather changes systematically with latitude. Some areas have been glaciated; other areas have not. Where weathering processes act to make poor soils poorer or rich soils richer, climatic and soil fertility asymmetries will align; where weathering makes poor soils richer or rich soils poorer, they will be antagonistic.

Asymmetry in land tenure regimes can have profound consequences for related social, economic, and political processes. Flying over the continental United States for the first time, one is immediately struck by the unusual symmetry in the landscape (figure 1.1).

The landscape resembles a quilt composed of square patches (one mile by one mile) bounded by county and state roads all running east-west or north-south. This symmetry reflects the fundamental principles on which the U.S. was built. The public land survey system that originally divided the public domain into these squares was designed by Thomas Jefferson around 1780. He believed that a democracy could only be built through dividing the land equally and creating a large middle class of yeoman farmers (Linklater 2002). Only through such symmetry would the social and economic asymmetries of feudal Europe be avoided. Where land distribution was and continues to be extremely asymmetrical, such as Latin America (figure 1.1b),

FIGURE 1.1. Two landscapes with very different levels of symmetry in land development patterns. (a) A relatively symmetrical landscape in the United States Midwest (Indiana). Note how, in this classification scheme, the different elements of this landscape are highly interchangeable; random rearrangements of patches would have little effect on its overall structure or function. (b) Asymmetrical land development patterns in Latin America (Bolivia).

there can be severe political, social, and economic consequences (Thiesen-husen 1995). The Jeffersonian ideals contrasted with the experience in Latin America illustrate the strong correlation between symmetry in land distribution, on the one hand, and political, social, and economic symmetry on the other hand.

Although many of the classical examples of landscape asymmetries are passive, in the sense that gradients are perceived to exist in the environment

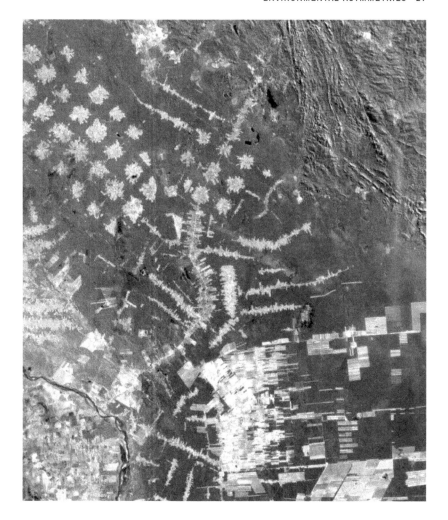

FIGURE 1.1. *continued.*

but to play a relatively static ecological role, asymmetries may be dynamic and variable. Dynamic analysis of landscape asymmetries is relatively recent in ecology, and there are few well-documented case studies to draw on in this context. In theory, however, some of the most interesting dynamics associated with asymmetries could occur if asymmetrical patterns and related processes are self-organizing through space and time. In this context, we take self-organization to mean that landscape patterns might influence the processes that produce or sustain them, resulting in a dynamic feedback between pattern and process. For example, fire is an essential component of vegetation dynamics in Alaska (Kasischke, Christiansen, and Stocks 1995). The different albedos (reflectance values) of spruce forests (fire susceptible) and tundra (fire sustained) in Alaska result in different temperature balances and air circulation patterns over forests and tundra respectively. Circulation patterns influence the formation of clouds and the incidence of thunderstorms. Thunderstorms in turn produce lightning, which is the main cause of fires in these areas. A situation thus exists in which the properties of the landscape can become self-maintaining; the fire-dependent, reflective tundra experiences a higher incidence of lightning strikes than the fire-susceptible, more absorptive spruce forest (Bonan, Chapin, and Thompson 1995; Higgins, Mastrandrea, and Schneider 2002; Kasischke, Christiansen, and Stocks 1995; Lafleur and Rouse 1995). Both habitat types influence the local disturbance regime in such a way as to maintain beneficial conditions for their own continuation. A similar kind of feedback between vegetation and fire, although with a different underlying mechanism, has been documented in Florida (Peterson 2002).

In other cases there will be feedbacks between different sets of asymmetries in the same environment. Feedbacks occur when an effect influences the magnitude or rate of its cause. For instance, asymmetries in fruit production can drive the movements of seed dispersing animals; and the movements of seed dispersers will influence the spatial distribution of fruit production. If fruit production and seed dispersal are both asymmetrical and enhance one another, then local amplifications of asymmetries may make existing environmental gradients steeper. In practice this dynamic could lead to a clustered distribution of fruiting trees and the formation of a fruit-based spatial network, with clusters of fruit trees as nodes and its edges defined by the pathways traveled by seed dispersing agents as they move along paths or flyways between nodes. Asymmetries in spatial patterns and plant-animal interactions will also occur at different scales within the same system (e.g., Jordano 1994; Olesen and Jordano 2002) and may interact with one another across scales, resulting in complex hierarchical dynamics. For example, rainfall and elevation asymmetries may influence seed dispersers directly through their effects on energy expenditure and ease of movement or indirectly via their effects on fruit and other resources.

In general, the potential for landscape self-organization (including, but not limited to, bifurcations and the formation of alternate stable states) arises

TABLE 1.1. Some Examples of Self-Organizing Environmental Asymmetries

PATTERN ASYMMETRY	SELF-ORGANIZING PROCESS OR MECHANISM.
Fuel load	Hotter fires in areas with higher fuel loads favor rapidly regenerating/pyrophilic species; heterogeneity begets further heterogeneity.
Edge effects	Edges facilitate edge-specific processes that create further edges, as in subdivision of properties or ramification of a road network.
Fruit production	Frugivory and seed dispersal by frugivores. High local abundance of fruit attracts more seed dispersers to region.
Frequency of lightning-ignited fires	Differing albedos between adjacent dark and light vegetation types result in differential, self-sustaining lightning and fire regimes.
Property sizes and associated income differences	Ownership of more resources generates greater wealth and allows purchase of more land.
Rainfall, transpiration, and vegetation	Higher rainfall leads to more vegetation, greater transpiration, water vapor, rainfall; reduced transpiration leads to reduced water vapor and rainfall.
Grazing lawns	Dung deposition favors production of high-nutrient grasses, which in turn attract large herbivores that deposit more dung.
Access to information	Individuals or firms who have access to more and better information have a competitive advantage over those lacking such information, which allows them to generate further information more easily than their competitors.
Elevation	Erosion. Steeper areas erode more rapidly, sediment is deposited in lower areas, the earth's surface becomes flatter, and asymmetry is reduced over time.

whenever asymmetries exist in a linked pattern-process relationship. This observation suggests that there may be a relationship between spatial and temporal variation. If the action of a given process varies in both space and time in response to asymmetries, then areas of higher spatial heterogeneity may be more variable in time than areas of lower spatial heterogeneity. In other words, habitats that show higher local heterogeneity may also exhibit faster turnover times in such things as vegetation types and nutrient cycling. To return to the fire example, areas in which fuel loads build up in a more variable manner will experience spatial differences in the intensity and duration of fires, resulting in an even more heterogeneous postfire landscape that may be prone to a second fire a few years later and a resulting shift in the patch mosaic of grassland, shrubs, and forest. By contrast, where fuel

FIGURE 1.2. Location of our study site in north central Florida. The black square indicates the footprint of the Landsat images used in the analysis.

loads are homogeneous, a single hot fire will reset a successional state in such a way that rates of landscape change will be slow for decades.

To the best of our knowledge, the question of causal feedbacks between spatial and temporal variation in landscapes has not been explicitly considered. We have recently undertaken a study of interactions between spatial and temporal asymmetries in the landscape of north central Florida (Southworth et al. 2006). We discuss this analysis in some detail because it provides an interesting example of how the interactions between spatial and temporal asymmetries can be considered across multiple scales.

The analysis used six Landsat images (1985, 1989, 1992, 1997, 2001, 2003) of a study area centered around Gainesville, Florida (figure 1.2). After applying standard processing methods, Normalized Difference Vegetation Index (NDVI) images (Jensen 1996) were created for each year to provide a continuous vegetation/land cover data set that varies in both space and time. NDVI is an index of primary production. For the analysis of asymmetries, the continuous nature of NDVI data offers an appealing alternative to the categorical data sets that are used in standard land cover maps (Southworth, Munroe, and Nagendra 2004). NDVI images were created for all image dates. To obtain images that showed the change through time at each location, we simply subtracted the earlier images from the later images. For example, subtraction of NDVI values for the year 2001 from NDVI values for the year 2003 produces a difference image for the period 2001 to 2003. Difference images were calculated for each of the lag times in the data set. We then used a moving window of different sizes (3×3, 10×10, 25×25, 50×50, 100×100, and 250×250 pixels) to calculate local spatial variation in both the original data and the difference images. The variance (sigma) of all NDVI values in the window is assigned to the central pixel; as the window moves through the data set, each location in the image is assigned a variance value.

To summarize, by this point in the analysis the imagery-derived measures represent the spatial variation in primary production (at a single time) and the change in spatial variation through time, at multiple spatial scales. The next step in the analysis is to regress our data for spatial variation (termed S for variation through Space) on our values for the change in spatial variation (termed T for spatial variation through Time) at multiple scales. If spatial and temporal variation are interacting in a landscape, we would expect to see a predictable relationship between S and T. For example, greater values of S would lead to greater values of T if pattern-process amplification occurs, resulting in a significant regression with a positive slope.

The regression analysis was undertaken for each window size and each pair of image dates. As a null model, we randomized the 1997 and 2001 images and repeated the analysis for these data and the difference image 2001–1997. The data were randomized in such a way as to maintain the same statistical properties as the original images while destroying any spatial asymmetries inherent in the data by rearranging the locations of data points. The values of the regression parameters for the S-T relationship change with scale, and there is no single 'correct' scale for this analysis, so a scaling component is integral to the study. Of particular interest was the question of whether there were predictable, scale-related changes in the slope and intercept of the regressions. Somewhat to our surprise, we found that several predictable relationships emerged. The strength of the relationship between spatial variance and temporal variance was parabolic, peaking

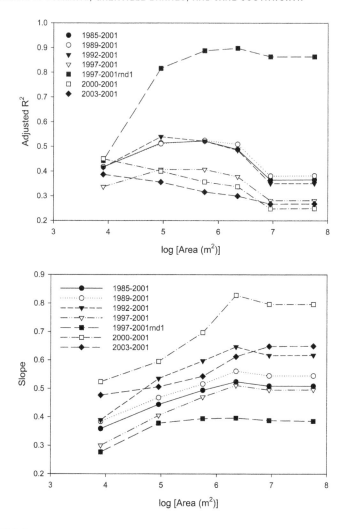

FIGURE 1.3. Scatterplots showing (a) the strength of the relationships and (b) the slope of a linear regression of temporal variation on spatial variation for a time series of NDVI data for north central Florida.

at a window size of 10x10 pixels (350 × 350m or 12.25ha; figure 1.3). The slope of the relationship between spatial variance and temporal variance was highest at a kernel size of 50x50 pixels (1750 × 1750m or 306.25ha). The ratio of temporal variance to spatial variance increased with the spatial scale of analysis. At smaller spatial scales, the spatial variation was greater than the temporal variation. With increasing spatial scale we also found increasing temporal variance. As the scale of analysis increased, the difference between the actual and random data increased, and so spatial variation explained less and less temporal variation than the null model (figure 1.3).

This analysis begs the question of mechanism, but it does demonstrate that spatial and temporal variation are closely related to one another in the landscape of northern Florida and that consideration of interacting asymmetries across multiple scales may ultimately allow us to derive relationships that have predictive power for landscape change. Many remote sensing studies are critiqued on the basis of a lack of transferability across space and time and a lack of consistent relationships (Foody and Atkinson 2003). This kind of research, in which general system properties such as heterogeneity and asymmetry are considered, has the potential to generate relationships that are both useful and transferable between different landscapes.

To what extent are the same methods applicable to the study of asymmetries in social systems? We applied our approach to a land tenure data set that showed ownership in a small subset of the same north Florida landscape. Ownership was recorded every five years over a thirty-year time horizon (figure 1.4; Barnes et al. [2003]).

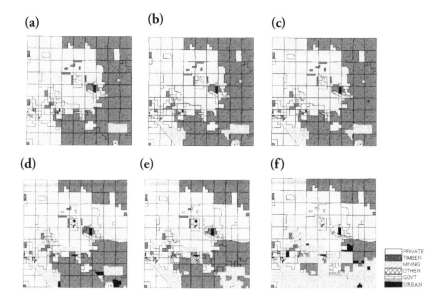

FIGURE 1.4. Maps showing changes in land ownership for Hamilton area. Each square (section) is approximately one square mile.

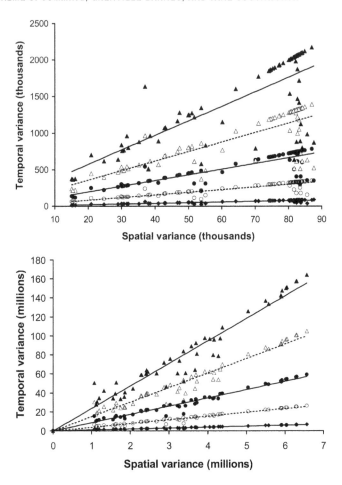

FIGURE 1.5. Plots showing the relationship between spatial and temporal variation in the Hamilton data in moving windows at different scales. (a) 1 × 1 section, (b) 3 × 3 sections, (c) 5 × 5 sections, (d) 7 × 7 sections. In each case the slope of the line increases with time relative to a 1974 baseline. Filled diamond, 1975–1980; hollow circle, 1975–1985; filled circle, 1970–1990; hollow triangle, 1975–1995; filled triangle, 1975–2000.

In this instance there were even stronger relationships between spatial and temporal variation, particularly over long time horizons (figures 1.5 and 1.6). The slope of the regression relationship becomes steeper over longer time horizons and shows no indication of leveling off, suggesting that the temporal scale at which land ownership changes is longer than the thirty-year scale of this data set. Taken in context, the results imply that changes in ownership are most likely to occur in areas where there are already multiple

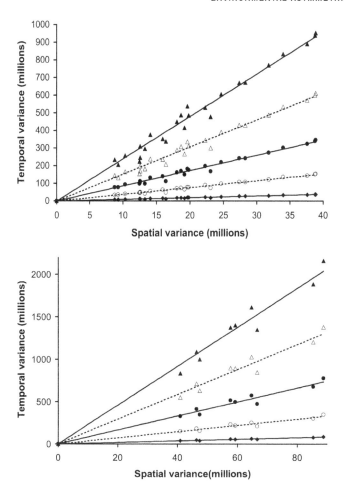

FIGURE 1.5. *continued.*

owners. These areas will often be at the edges of existing properties. This conclusion makes sense in light of the mechanisms that are likely to lead to subdivision of a property; it is considerably more likely that someone will sell land that is at the periphery of their existing plot than in the center, and it is also more likely that an owner who is interested in expanding their property in a rural setting will focus their attention on purchasing blocks of land that are adjacent to their current holdings.

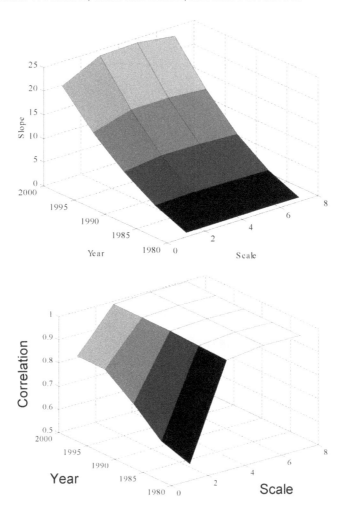

FIGURE 1.6. Three-dimensional plots showing the relationship between scale, time step relative to a 1975 baseline, and (a) slope of the linear regression of temporal variation on spatial variation or (b) the strength of the correlation between spatial and temporal variation, as measured by the r^2 statistic.

Land tenure data are often considered using GINI indices. The classical GINI index is defined as

$$G = \frac{\sum_{i=1}^{n} \sum_{j=1}^{n} \left| x_i^n - x_j \right|}{2n^2 \bar{x}}$$

where x is an observed value, n is the number of values observed, and x bar is the mean value of all x. The GINI index is a simple measure of asymmetry for a particular data set. It has been widely used in national-level analyses of poverty and human health. The index in formula 1 is often depicted as a Lorenz curve, in which the cumulative proportional increase in income (or other quantity of interest) per person, sorted from smallest to largest, is plotted against the number of people. A straight diagonal line ($x = y$) for the Lorenz curve indicates a completely equitable distribution of resources, because each additional person adds an equivalent amount of wealth, health, or land. The area under the Lorenz curve is the GINI index, which ranges from 0 to 1 and is highest when equity is lowest. In this particular example we considered GINI indices for the areas of different parcel sizes. Land parcels were grouped together into one of six different ownership types (private, timber, mining, government, urban, other; see figure 1.5) at each of five different scales (subparcel, parcel, 3 × 3 parcels, 5 × 5 parcels, and 7 × 7 parcels), and GINI indices were calculated for the resulting area data. Although there is little change in the GINI index for the study area over the time period of this study, the scale-dependency of the GINI index is quite obvious (figure 1.7). If land ownership were a contentious issues in this system, we could easily envisage that asymmetries in parcel size could drive social or economic processes (such as stock theft, the formation of lobby or special interest groups, or price wars in the local market) at some scales but not at others, depending on the alignment between the scales at which land tenure occurs and the scales at which socioeconomic processes occur.

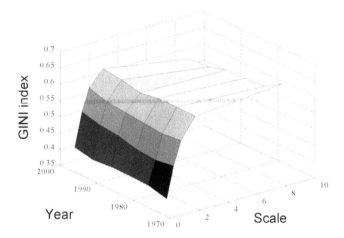

FIGURE 1.7. Three-dimensional plot showing the scale dependency (x) of the GINI index (z), a simple measure of asymmetry, at different time steps (y).

GENERAL DISCUSSION AND CONCLUSIONS

Asymmetries in landscapes arise from a number of different processes across a wide range of different scales. Once asymmetries have arisen, they may either facilitate their own continuation or may gradually disappear. Many kinds of landscape asymmetry have arisen from slowly-acting, broad-scale processes and patterns such as continental drift and the size and shape of the earth. Other kinds of landscape asymmetry have been generated relatively recently by organisms and their interactions with their environment. Landscapes provide the context for evolution, and organisms have shaped their environments as they have evolved (Levin 1999). More recently, and more rapidly, humans have had a huge and disproportionate effect on landscape pattern and process. Each of these different drivers of change has created its own set of feedbacks and influences on the arrangement of heterogeneity within landscapes. To fully understand the consequences of landscape asymmetries, we thus need to understand not only the ways in which organisms and human societies respond to slowly changing asymmetries (such as elevation and rainfall gradients or cultural norms) but also the different kinds of feedback that serve to stabilize system dynamics or propel them along novel trajectories.

Levin (1999) argues that some kind of selection process, by which individual system components are removed or enhanced, is a central attribute of complex adaptive systems. Landscape asymmetries can be considered as gradients that serve to drive such processes, as discussed in later chapters in this volume. However, it is important to note that asymmetries themselves may also be the subject of selection or enhancement by both internal and external processes. Although the broader landscape can be viewed as a passive and fairly constant context in which other ecological and sociological processes occur, it is also in constant flux at one scale or more.

Research on the role of landscape asymmetries in CASs is currently in its infancy. Most ecological research has focused on organismal responses to asymmetries. A small but growing body of research exists on the dynamics of complex systems, particularly in regard to the broadscale feedbacks between different kinds of system components (e.g., Higgins, Mastrandrea, and Schneider 2002; Scheffer et al. 2001). One of the steps that will be necessary to relate this research to environmental asymmetries is to make it more explicitly spatial so that we can understand the context dependence of different kinds of environment-organism feedbacks. Further analysis of land tenure regimes and how these change through time is also a fertile area for expanding our understanding of complex social-ecological systems, as is the study of self-organization and the mechanisms that lead to landscape change. The growing fields of metapopulation and metacommunity analysis may contribute usefully to this agenda, particularly if synergies can be found between metacommunity studies, land use/land cover studies, and studies of the higher-level properties of landscapes (such as landscape diversity, permeability, and fractal dimension). It seems to us that the concept of environmental asymmetry, as defined in this chapter, has much to offer the further theoretical and empirical development of complex systems theory and its applications to real-world problems.

REFERENCES

Acemoglu, D., and A. Scott. 1997. "Asymmetric Business Cycles: Theory and Time-Series Evidence." *Journal of Monetary Economics* 40:501–533.

Ahtiainen, J.J., R.V. Alataio, J. Mappes, and L. Vertainen. 2003. "Fluctuating Asymmetry and Sexual Performance in the Drumming Wolf Spider Hygrolycosa Rubrofasciata." *Annales Zoologici Fennici* 40:281–292.

Allen, A.P., J.H. Brown, and J.F. Gillooly. 2002. "Global Biodiversity, Biochemical Kinetics, and the Energetic-Equivalence Rule." *Science* 297:1545–1548.

Amir, R., and J. Wooders. 1998. "Cooperation vs. Competition in R and D: The role of stability of Equilibrium." *Journal of Economics-Zeitschrift Fur Nationalokonomie* 67:63–73.

Bakun, A., and K. Broad. 2003. "Environmental 'Loopholes' and Fish Population Dynamics: Comparative Pattern Recognition with Focus on El Nino Effects in the Pacific." *Fisheries Oceanography* 12:458–473.

Barnes, G., A. Agrawal, L. Genc, B. Ramachandran, V. Sivaraman, B. Pudi, M. Binford, and S. Smith. 2003. "Developing a Spatio-temporal Cadastral Database Using County Appraisal Data from Northern Florida." *Surveying and Land Information Science* (special cadastral issue) 63:243–251.

Berger, J. 2004. "The Last Mile: How to Sustain Long-Distance Migration in Mammals." *Conservation Biology* 18:320–331.

Best, P.B., and D.M. Schell. 1996. "Stable Isotopes in Southern Right Whale (Eubalaena Australis) Baleen as Indicators of Seasonal Movements, Feeding and Growth." *Marine Biology* 124:483–494.

Bonan, G.B., F.S. Chapin, and S.L. Thompson. 1995. "Boreal Forest and Tundra Ecosystems as Components of the Climate System." *Climatic Change* 29:145–167.

Borisova, I.A., and V.G. Kazmin. 1993. "Dependence of the Ocean Depth on the Age of the Sea-Floor in the North-Atlantic in Connection with History of Continental Break-Up." *Okeanologiya* 33:915–923.

Brading, K., and E. Castellani. 2003. "Introduction." In K. Brading, and E. Castellani, eds., *Symmetries in Physics: Philosophical Reflections*, 1–18. Cambridge: Cambridge University Press.

Brown, C.T., and W.R.T. Witschey. 2003. "The Fractal Geometry of Ancient Maya Settlement." *Journal of Archaeological Science* 30:1619–1632.

Brown, J.H. 1995. *Macroecology*. Chicago: University of Chicago Press.

Buyanovsky, A.I. 1996. "Bimodality and Asymmetry in Size-Frequency Distribution of the Young Mussels Mytilus Trossulus (Bivalvia, Mytilidae)." *Zoologichesky Zhurnal* 75:978–984.

Cadenasso, M.L., S.T.A. Pickett, K.C. Weathers, S.S. Bell, T.L. Benning, M.M. Carreiro, and T.E. Dawson. 2003. "An Interdisciplinary and Synthetic Approach to Ecological Boundaries." *BioScience* 53:717–722.

Castellani, E. 2003. "On the Meaning of Symmetry Breaking." In K. Brading and E. Castellani, eds., *Symmetries in Physics: Philosophical Reflections*, pp. 321–334. Cambridge: Cambridge University Press.

Chen, Y.G., and Y.X. Zhou. 2003. "The Rank-Size Rule and Fractal Hierarchies of Cities: Mathematical Models and Empirical Analyses." *Environment and Planning B—Planning and Design* 30:799–818.

Coe, M.J., D.H. Cumming, and J. Phillipson. 1976. "Biomass and Production of Large African Herbivores in Relation to Rainfall and Primary Production." *Oecologia* 22:341–354.

Cooper, M., D.H. Downs, and G.A. Patterson. 2000. "Asymmetric Information and the Predictability of Real Estate Returns." *Journal of Real Estate Finance and Economics* 20:225–244.

Cumming, D.H.M., and G.S. Cumming. 2003. "Ungulate Community Structure and Ecological Processes: Body Size, Hoof Area and Trampling in African Savannas." *Oecologia* 134:560–568.

Cummings, V., A. Jones, and A. Watson. 2002. "Divided Places: Phenomenology and Asymmetry in the Monuments of the Black Mountains, Southeast Wales." *Cambridge Archaeological Journal* 12:57–70.

Dargay, J., and D. Gately. 1995. "The Response of World-Energy and Oil Demand to Income Growth and Changes in Oil Prices." *Annual Review of Energy and the Environment* 20:145–178.

de Vries, H., and J.C. Biesmeijer. 2002. "Self-organization in Collective Honeybee Foraging: Emergence of Symmetry Breaking, Cross Inhibition and Equal Harvest-Rate Distribution." *Behavioral Ecology and Sociobiology* 51:557–569.

Deininger, K. 2003. *Land Policies for Growth and Poverty Reduction*. Washington, DC: World Bank and Oxford University Press

Despland, E., J. Rosenberg, and S.J. Simpson. 2004. "Landscape Structure and Locust Swarming: A Satellite's Eye View." *Ecography* 27:381–391.

Du Toit, J. T. 1995. "Determinants of the Composition and Distribution of Wildlife Communities in Southern Africa." *Ambio* 24:2–6.

Foody, G., and P. Atkinson 2003. *Uncertainty in Remote Sensing and GIS*. London: Wiley.

Fryxell, J. M., J. F. Wilmshurst, and A. R. E. Sinclair. 2004. "Predictive Models of Movement by Serengeti Grazers." *Ecology* 85:2429–2435.

Gruzelier, J. H., L. Wilson, D. Liddiard, E. Peters, and L. Pusavat. 1999. "Cognitive Asymmetry Patterns in Schizophrenia: Active and Withdrawn Syndromes and Sex Differences as Moderators." *Schizophrenia Bulletin* 25:349–362.

Higgins, P. A. T., M. D. Mastrandrea, and S. H. Schneider. 2002. "Dynamics of Climate and Ecosystem Coupling: Abrupt Changes and Multiple Equilibria." *Philosophical Transactions of the Royal Society: Biological Sciences* 357: 647–655.

HilleRisLambers, R., M. Rietkirk, F. Van Den Bosch, H. H. T. Prins, and H. De Kroon. 2001. "Vegetation Pattern Formation in Semi-arid Grazing Systems." *Ecology* 82:50–61.

Jensen, J. R. 1996. *Introductory Digital Image Processing: A Remote Sensing Perspective*. Saddle River, NJ: Prentice-Hall.

Jordano, P. 1994. "Spatial and Temporal Variation in the Avian-Frugivore Assemblage of *Prunus Mahaleb*: Patterns and Consequences." *Oikos* 71:479–491.

Kasischke, E. S., N. L. Christiansen, and B. J. Stocks. 1995. "Fire, Global Warming, and Carbon Balance of Boreal Forests." *Ecological Applications* 5:437–451.

Keating, B. A., D. Gaydon, N. I. Huth, M. E. Probert, K. Verburg, C. J. Smith, and W. Bond. 2002. "Use of Modelling to Explore the Water Balance of Dryland Farming Systems in the Murray-Darling Basin, Australia." *European Journal of Agronomy* 18:159–169.

Lafleur, P. M., and W. R. Rouse. 1995. "Energy Partitioning at Treeline Forest and Tundra Sites and Its Sensitivity to Climate Change." *Atmosphere-Ocean* 33:121–133.

Lemoine, B. 1986. "Symmetry, Form and Character in Architecture—French—Szambien,W." *Architecture D'Aujourdhui* 246:R47–R47.

Levin, S. A. 1976. "Spatial Patterning and the Structure of Ecological Communities," pp. 1–36. In S. A. Levin, ed., *Lectures on Mathematics in the Life Sciences*. Vol. 8: *Some Mathematical Questions in Biology VII*. Providence, R. I., American Mathematical Society.

Levin, S. A. 1999. *Fragile Dominion: Complexity and the Commons*. Cambridge: Perseus.

Linklater, A. 2002. *Measuring America: How an Untamed Wilderness Shaped the United States and Fulfilled the Promise of Democracy*. New York: Walker.

Logli, F., and R. Joffre. 2001. "Individual Variability as Related to Stand Structure and Soil Condition in a Mediterranean Oak Coppice." *Forest Ecology and Management* 142:53–63.

Lomolino, M. V. 2001. "Elevation Gradients of Species-Density: Historical and Prospective Views." *Global Ecology and Biogeography* 10:3–13.

Machtans, C. S., M.-A.Villard, and S. J. Hannon. 1996. "Use of Riparian Buffer Strips as Movement Corridors by Forest Birds." *Conservation Biology* 10:1366–1379.

MDBC. 1999. *The Salinity Audit of the Murray-Darling Basin.* Canberra: Murray-Darling Basin Commission. http://www.mdbc.gov.au/salinity/salinity_audit_1999.

Meyburg, B. U., P. Paillat, and C. Meyburg. 2003. "Migration Routes of Steppe Eagles Between Asia and Africa: A Study by Means of Satellite Telemetry." *Condor* 105:219–227.

Middleton, R. 1989. "Symmetry, Taste, Character: Theory and Terminology of Classical-Age Architecture, 1500–1800." *Burlington Magazine* 131:44–45.

Miller, D. 2003. "An Asymmetry-Based View of Advantage: Towards an Attainable Sustainability." *Strategic Management Journal* 24:961–976.

Olesen, J., and P. Jordano. 2002. "Geographic Patterns in Plant-Pollinator Mutualistic Networks." *Ecology* 83:2416–2424.

Palmer, T. M. 2003. "Spatial Habitat Heterogeneity Influences Competition and Coexistence in an African Acacia Ant Guild." *Ecology* 84:2843–2855.

Peters, R. H. 1983. *The Ecological Implications of Body Size.* Cambridge: Cambridge University Press.

Peterson, G. D. 2002. "Estimating Resilience Across Landscapes." *Conservation Ecology* 6:17. http://www.consecol.org/vol16/iss11/art17.

Polis, G. A., M. E. Power, and G. R. Huxel 2004. *Food Webs at the Landscape Scale.* Chicago: University of Chicago Press.

Portha, S., J. L. Deneubourg, and C. Detrain. 2002. "Self-organized Asymmetries in Ant Foraging: A Functional Response to Food Type and Colony Needs." *Behavioral Ecology* 13:776–781.

Pushcharovsky, Y. M., and Y. P. Neprochnov. 2003. "Tectonics and Deep Structure of Deep-Sea Basins in the North of the Central Atlantic Ocean." *Geotectonics* 37:95–106.

Rajaniemi, T. K. 2003. "Evidence for Size Asymmetry of Belowground Competition." *Basic and Applied Ecology* 4:239–247.

Richter, H. V., and G. S. Cumming. 2006. "Food Availability and the Annual Migration of the Straw-Coloured Fruit Bat (*Eidolon helvum*) at Kasanka National Park, Zambia." *Journal of Zoology, London* 268:35–44.

Rietkirk, M., P. Ketner, J. Burger, B. Hoorens, and H. Olff. 2000. "Multiscale Soil and Vegetation Patchiness Along a Gradient of Herbivore Impact in a Semi-arid Grazing System in West Africa." *Plant Ecology* 148:207–224.

Scheffer, M., S. R. Carpenter, J. A. Foley, C. Folke, and B. Walker. 2001. "Catastrophic Shifts in Ecosystems." *Nature* 413:591–596.

Schreiner, W., F. Neumann, M. Neumann, R. Karch, A. End, and S. M. Roedler. 1997. "Limited Bifurcation Asymmetry in Coronary Arterial Tree Models

Generated by Constrained Constructive Optimization." *Journal of General Physiology* 109:129–140.

Schwinning, S., and J. Weiner. 1998. "Mechanisms Determining the Degree of Size Asymmetry in Competition Among Plants." *Oecologia* 113:447–455.

Simon, H. A. 1962. "The Architecture of Complexity." *Proceedings of the American Philosophical Society* 106:467–482.

Southworth, J., G. S. Cumming, M. Marsik, and M. Binford. 2006. "Linking Spatial and Temporal Variation at Multiple Scales in a Heterogeneous Landscape." *Annals of Geography* 58:406–420.

Southworth, J., D. Munroe, and H. Nagendra. 2004. "Land Cover Change and Landscape Fragmentation—Comparing the Utility of Continuous and Discrete Analyses for a Western Honduras Region." *Agriculture, Ecosystems and Environment* 101:185–205.

Stewart, I. 2003. "Self-organization in Evolution: A Mathematical Perspective." *Philosophical Transactions of the Royal Society of London, Series A* 361: 1101–1123.

Tarboton, D. G., R. L. Bras, and I. Rodriguez-Iturbe. 1988. "The Fractal Nature of River Networks." *Water Resources Research* 24:1317–1322.

Thiesenhusen, W. 1995. *Broken Promises: Agrarian Reform and the Latin American Campesino.* Boulder: Westview.

Thirgood, S., A. Mosser, S. Tham, G. Hopcraft, E. Mwangomo, T. Mlengeya, M. Kilewo, J. Fryxell, A. R. E. Sinclair, and M. Borner. 2004. "Can Parks Protect Migratory Ungulates? The Case of the Serengeti Wildebeest." *Animal Conservation* 7:113–120.

Thomas, D. W. 1983. "The Annual Migrations of Three Species of West-African Fruit Bats (Chiroptera, Pteropodidae)." *Canadian Journal of Zoology* 61:2266–2272.

Toffler, A. 1980. *The Third Wave.* New York: Morrow.

Toomey, D. R., W. S. D. Wilcock, J. A. Conder, D. W. Forsyth, J. D. Blundy, E. M. Parmentier, and W. C. Hammond. 2002. "Asymmetric Mantle Dynamics in the MELT Region of the East Pacific Rise." *Earth and Planetary Science Letters* 200:287–295.

Trinkel, M., P. H. Fleischmann, A. F. Steindorfer, and G. Kastberger. 2004. "Spotted Hyenas (Crocuta Crocuta) Follow Migratory Prey. Seasonal Expansion of a Clan Territory in Etosha, Namibia." *Journal of Zoology* 264:125–133.

Turner, M. G., R. H. Gardner, and R. V. O'Neill 2001. *Landscape Ecology in Theory and Practice: Pattern and Process.* New York: Springer.

van Beek, J. 1997. "Is Local Metabolism the Basis of the Fractal Vascular Structure in the Heart?" *International Journal of Microcirculation-Clinical and Experimental* 17:337–345.

Van Dongen, S., L. Lens, and G. Molenberghs. 1999. "Mixture Analysis of Asymmetry: Modelling Directional Asymmetry, Antisymmetry and Heterogeneity in Fluctuating Asymmetry." *Ecology Letters* 2:387–396.

vanderMeer, P. J., and F. Bongers. 1996. "Formation and Closure of Canopy Gaps in the Rain Forest at Nouragues, French Guiana." *Vegetatio* 126:167–179.

Vannote, R. L., G. W. Minshall, K. W. Cummins, J. R. Sedell, and C. E. Cushing. 1980. "The River Continuum Concept." *Canadian Journal of Fisheries and Aquatic Sciences* 37:130–137.

Vollestad, L. A., K. Hindar, and A. P. Moller. 1999. "A Meta-analysis of Fluctuating Asymmetry in Relation to Heterozygosity." *Heredity* 83:206–218.

Walther, B. A., M. S. Wisz, and C. Rahbek. 2004. "Known and Predicted African Winter Distributions and Habitat Use of the Endangered Basra Reed Warbler (Acrocephalus Griseldis) and the Near-Threatened Cinereous Bunting (Emberiza Cineracea)." *Journal of Ornithology* 145:287–299.

Ward, T. M., J. Staunton-Smith, S. Hoyle, and I. A. Halliday. 2003. "Spawning Patterns of Four Species of Predominantly Temperate Pelagic Fishes in the Subtropical Waters of Southern Queensland." *Estuarine Coastal and Shelf Science* 56:1125–1140.

Wedel, M., W. A. Kamakura, W. S. Desarbo, and F. Terhofstede. 1995. "Implications for Asymmetry, Nonproportionality, and Heterogeneity in Brand Switching from Piece-Wise Exponential Mixture Hazard Models." *Journal of Marketing Research* 32:457–462.

White, R., and G. Engelen. 1993. "Cellular-Automata and Fractal Urban Form—A Cellular Modeling Approach to the Evolution of Urban Land-Use Patterns." *Environment and Planning* A 25:1175–1199.

Whittaker, R. H. 1960. "Vegetation of the Siskiyou Mountains, Oregon and California." *Ecological Monographs* 30:279–338.

Wilson Jones, M. 2001. "Doric Measure And Architectural Design 2: A Modular Reading of the Classical Temples." *American Journal of Archaeology* 105:675–713.

Yizhaq, H., B. A. Portnov, and E. Meron. 2004. A mathematical model of segregation patterns in residential neighbourhoods. *Environment and Planning* A 36:149–172.

2

DIVERSITY AND RESILIENCE
OF SOCIAL-ECOLOGICAL SYSTEMS

Jon Norberg, James Wilson, Brian Walker, and Elinor Ostrom

OPTIONS OR ALTERNATIVES are fundamental requirements for change. Without them there can be no change or learning. Options and alternatives, however, come at a cost. The presence of too many alternatives may decrease efficiency through the expense of maintaining them, their influence on transaction or information-processing costs. Under a worldview of short-term profit maximization and a predictable or stable future, such costs are seen as unnecessary and easily become the target of sanitation efforts to increase short-term efficiency. As a consequence, options and alternatives are being lost globally in such ecosystem components as species and genotypes (FAO 1993), as well as in such social system components as languages, institutions (Ostrom 2005), local knowledge and information, and sets of actions that agents might take (Fardon 1995; Folke, Colding, and Berkes 2003). However, there is a growing recognition that diversity is a key requirement for long-term (sustainable) functioning of systems—biological and social (Low et al. 2003). We will demonstrate here that there are many fundamental reasons why diversity matters that are general enough to apply to two very different settings: species in ecosystems and institutions in governance. Our aim is to identify those aspects of diversity that are important, why they are important, how they matter for the system, and how they can be managed.

Contemporary theory on complex adaptive systems recognizes that some processes are needed to sustain diversity in order to maintain options for change over time (Schmalensee 1972; Levin 1998; Ostrom 1998, Edelman and Gally 2001). Diversity itself is, therefore, a dynamic attribute, and understanding the processes that alter and maintain diversity is an important step in defining the possibilities of creating sustainable social-ecological systems. Adaptive capacity is a process by which the relative abundance of system components changes in relation to how well they perform. In natural systems this feedback is reactive and autonomous, that is, dominant competitors become more abundant over time. Under changing environmental conditions the relative abundance of species shift such that system function is maintained. In social systems this feedback may be proactive (i.e., designed)—more or less influenced by the human capacity for information processing, planning, and foresight (Anderies and Norberg, chapter 5, this volume), and it amounts to the capacity of people in the system to manage the system's resilience. The consequences of a loss of diversity for the resilience of coupled social-ecological systems depends on the context and the attributes of the system elements that are lost.

An important concept for the study of social-ecological systems is resilience, originally formulated for describing the principal dynamics of ecosystems (Holling 1973). Many systems, ecological, social, and social-ecological, have multiple inherent feedback processes such that they experience several self-reinforcing states of attractions, called regimes (Walker and Meyers 2004). Shifts between such regimes can occur when the system is pushed across some threshold by external or internal dynamics (Scheffer and Carpenter 2003, Folke et al. 2004). Note that the definition of regimes differs from how it is used in the social sciences, where regime frequently refers to the political actors or system in power. Resilience is defined as the capacity of a system to absorb disturbance and reorganize while undergoing change so as to retain essentially the same function, structure, identity and feedbacks, that is, remain within one regime (Walker et al. 2004). Hence, resilience reflects the degree to which a complex adaptive system is capable of self-organization (versus lack of organization or organization forced by external factors) and the degree to which the system can build and sustain the capacity for learning and adaptation (Carpenter et al. 2001; Levin 1999). We can here define adaptive capacity (adaptability, *sensu* Walker et al. 2004) as the capacity of the system to alter the relative abundance of its components without significant changes in crucial system functions. Diversity thus plays a central role for resilience.

Since the notions of redundancy and diversity are often confused, we will here define our usage: assume we have a group of units and that each unit is characterized by ten attributes. If all units have the same values for each attribute, we have "true" redundancy. If there are differences in values in one or another attribute we have diversity. If one considers some important process that is determined by a given attribute, we can have redundancy in that attribute by means of diversity in other attributes. The deliberate inclusion of redundancy in airplane control systems provide an easy way to explain this. We want redundancy in controllability (i.e., the systems all need to do the same thing), but we don't want identical systems in all attributes since this means they would be equally vulnerable to external shock (e.g., a rapid change in temperature) or internal dynamics (e.g., failure of the hydraulic system). Thus the control systems are localized in different places on the airplane and might use slightly different mechanisms, such as mechanic versus hydraulic controls.

In this book we are most interested in understanding the dynamics of social-ecological systems and the problems associated with management of such systems (see also Anderies, Janssen, and Ostrom 2004). The time scales of interest are on the order of years to centuries at most and the level of interaction between human society and nature is at the level of ecosystems rather than single populations. Evolutionary processes within species are not our major focus here. Rather, the focal scale is the ecosystem, with local populations as its smallest unit, and local governance systems, with local rules as its smallest unit. While we recognize that understanding the whole cannot be achieved by reducing it to its elementary parts, these parts can still be analyzed as parts of interacting subsystems (Koestler 1967). There are many other types of subsystems involved in social-ecological systems, but these two are particularly relevant in terms of the performance of linked social-ecological systems as a whole (Folke et al. 2004). While our perspective is that of the entire system, the focal scale of analyses in this chapter will be on the ecosystem and the governance system.

THE ROLE OF BIODIVERSITY IN THE RESILIENCE OF ECOSYSTEM SERVICES

Ecosystem services are generated by structures or processes sustained by one or several species. Species that perform similar ecosystem functions or have similar structures are defined as functional groups. Some functional groups are directly related to ecosystem services that humans easily can appreciate, such as whales for tourism or forests trees for timber. Other functional groups of species, such as bacteria, provide the support system such as nutrient regeneration, waste breakdown, and production of gas. These are the "unsung heroes" of the more conspicuous ecosystem service providers (Daily 1997). The performance of functional groups of species may thus affect ecosystem services directly or indirectly. Furthermore, interactions among functional groups of species are often nonlinear responses to the abundances of other species (Holling 1959a, 1959b). Such nonlinearities may introduce thresholds that can be crossed if some functional group experiences dramatic decline in performance, e.g., due to some disturbance (see review in Folke et al. 2004; Walker and Meyers 2004, RA database on thresholds). The performance of functional groups is thus crucial for sustaining ecosystem services in changing natural environments; diversity within functional groups largely determines their ecosystem performance.

FUNCTION AND ATTRIBUTES OF SPECIES

Species and individuals are distinguished by their morphological and physiological traits. In terms of diversity, however, the focus in ecology has historically not emphasized these traits but has been concerned only with the numbers and abundance of species. For example, typical measures of species diversity, such as the Shannon-Wiener index, are based on the relative abundance of species and are not concerned with what these species actually do in the environment. However, given the recent focus in ecological research on the role of species diversity for ecosystem functioning, it is clear that to understand the role of diversity for ecosystem functioning one needs to focus on the diversity of traits in a species community, rather than on species per se (Grime 1997; Walker, Kinzig, and Langridge 1999; Loreau et al. 2001; Heemsbergen et al. 2004). This has the advantage that trait diversity, as opposed to species richness, can be mechanistically related to ecosystem functioning (Norberg et al. 2001; Norberg 2004). Species richness may only be indirectly related to ecosystem functioning if increasing the number of species in general also increases the trait that is crucial for any important ecosystem process (Petchey and Gaston 2002). In recent literature functional diversity has been assessed by means of quantitative traits (e.g., Walker, Kinzig, and Langridge 1999; Diaz and Cabido 2001; Petchey and Gaston 2002; Petchey, Hector, and Gaston 2004; Norberg 2004).

There are three general attributes of species that affect their ecological performance: 1. what species are and do (morphology such as size, biomass and structure and physiology such as movement, excretion, spawning),

2. how they respond (sensitivity of morphology and physiology to external drivers), and 3. whom they interact with (the resources and predators, i.e., the food web). The terminology used to describe these different traits and the mechanisms by which they affect ecosystem functioning has not been very consistent or well defined (Tilman, Lehman, and Thomson 1997; Diaz and Cabido 2001; Nystrŝm and Folke 2001; Petchey and Gaston 2002; Elmqvist et al. 2003; Hooper et al. 2005). Especially confusing is the use of functional groups, functional diversity, and ecosystem functioning. At times, the "function" is the same for all three terms, but it can also mean different things. For example, coral communities consist of three functional groups, i.e., massive, branching, and foliaceous species. Functional diversity may refer to their mode of dispersal, and ecosystem functioning may refer to their contribution to primary production. Furthermore, functional groups are often defined by their resource use, such as filter feeders versus deposit feeders, but sometimes other attributes such as morphology, dispersal, or behavior are used, e.g., massive versus branching corals, or long distance spawners versus short distance spawners (Steneck 2001). An approach less prone to confusion is to avoid the term *ecosystem functioning* and instead refer directly to the ecosystem process (e.g., primary production, nutrient regeneration) or structure (e.g., standing biomass, structure in coral reefs) that provides an important ecosystem service (*sensu* Daily 1997). An additional benefit of this is that a value can be attributed to an ecosystem service, in terms of being better or worse from a human or social perspective, and thus be relevant to management. An ecosystem process is of course not better or worse in itself, but only through the values that human society applies to it.

The interesting question then is how the diversity in attributes of all species that contribute to a particular ecosystem service affects their combined efficiency in sustaining it. Different mechanisms affect the performance of such potentially diverse groups of species. First, if species respond differently to changing conditions, fluctuations will have less effect on the performance of the group as a whole and thus on the ecosystem service (Walker, Kinzig, and Langridge 1999; Norberg et al. 2001; Hughes et al. 2003; Bai et al. 2004). The diversity of species within a functional group is called response diversity (Elmqvist et al. 2003). Second, if species are to some degree complementary in the resource they use, the performance of the group as a whole can increase when the degree of resource use, or the potential increase in biomass as a result thereof, is correlated with the ecosystem service provided by these species. We call this effect functional diversity, but it has added complexity as described in the next section. If there is no complementarity, competition will often result in competitive exclusion of all but one species. Thus the degree of complementarity is usually negatively correlated with the degree of competition.

FUNCTIONAL DIVERSITY

Functional diversity generally has been used to relate differences in morphological and physiological traits, response traits, and resource use traits of species (Walker, Kinzig, and Langridge 1999; Petchey and Gaston 2002; Naeem and Wright 2003; Folke et al. 2004). Naeem and Wright (2003) and others (e.g., Hooper et al. 2005) argue for considering response diversity as a separate measure. Functional diversity is most often used to mean complementarity in resource use and is sometimes equivalent to functional groups. But even within functional groups, a degree of complementary in resource use often exists. It might therefore be more consistent to think of the complementary aspect of functional diversity as a general measure, and functional group diversity as a measure of functional diversity with coarser resolution. Functional diversity not only addresses parallel interactions, such as use of different resources by species within the same trophic level, but may also include serial interactions such as trophic chains or particular morphologies or behaviors such as seed dispersers. There are now established measures of functional diversity (Walker, Kinzig, and Langridge 1999; Petchey and Gaston 2002; Petchey, Hector, and Gaston 2004) that involve the summed differences in attributes. But different attributes may affect the efficiency by which an ecosystem service is affected in different ways. For example, the complementarity aspect often is positively correlated to production and biomass of a group of plant species, while the seed-dispersing aspect of functional diversity does not necessarily have any large effect on plant production but may be crucial for reestablishment after disturbances (Elmqvist et al. 2003).

Complementarity entails species having partially overlapping contributions to ecosystem function (and therefore partially overlapping resource requirements; Tilman, Lehman, and Thomson 1997). If two species use different resources, they can potentially coexist because one will not deplete the other's fundamental resources below its minimal requirement (Tilman

1982). If the degree of complementarity is large enough, coexistence can occur. Ecological theory on limiting similarity deals explicitly with the degree of complementarity that is necessary for coexistence (MacArthur and Levins 1967). A consequence of complementarity in essential resources between two species is that together they use more resources than either species alone. This may cause higher production and potentially higher biomass. The phenomenon has been referred to as overyielding (Loreau and Hector 2001). Complementarity in responses has similar effects (Hooper 1998) but is here measured over time, as we discuss in the next section. When two species share the same resource, some degree of competition can occur if the resource becomes limiting. The degree of competition depends on the degree of resource limitation and the degree of resource overlap. Such resource overlap can, for example, depend on attributes such as root depth or spatial location (Pacala and Levin 1997). The minimum requirement for coexistence to sustain a group of ecologically identical (neutral) species is the spatial dimension, i.e., attributes of locality (Pacala and Levin 1997; Hubbel 2001). Note that spatial separation is one form of resource partitioning that can cause complementarity effects. Spatiotemporal variability is a fundamental cause of diversity since, on a regional scale, it creates a multitude of opportunities for species with different attributes to coexist (Chesson 2000).

Functional effects on ecosystem properties depend on the interaction between species and their resources. Such interactions provide the network of the food web (see Webb and Bodin, chapter 3, this volume; Dunne, Williams, and Martinez 2002). However, network links can be serial, as in predator-prey, or parallel, as in species competing for the same resource, and this difference has large consequences for ecosystem functioning (Hulot 2000). While parallel interactions determine the degree of competition or complementarity as outlined above, serial interactions, like trophic chains, increase interdependence and give rise to complex dynamics (Hairston, Smith, and Slobodkin 1960), especially if the response of the species to the interaction strength is nonlinear (Brown et al. 2001). Such interactions can have strong effects on system structure and function and make the system inherently more sensitive to the loss of local populations such as evidenced in trophic cascades (Pace et al. 1999), keystone species (Paine 1969), or ecosystem engineers (Jones, Lawton, and Shachak 1994). One can think of functional diversity as a measure of the diversity of links from the species in the functional group to their resources or consumers rather than the number of species in the group. Thus, both effects of complementarity and of predator-prey relationships are incorporated in functional diversity as well as positive direct interactions such as facilitation (e.g., Holmgren et al. 1997). Functional diversity therefore encompasses all interactions between the species that make up the network, i.e., the food web, including competition, complementarity, and facilitation within trophic levels. Complementarity in resource use is the aspect of functional diversity that has received most focus.

RESPONSE DIVERSITY AND SPATIAL RESILIENCE

Diaz and Cabido (2001) and Hooper et al. (2005) use the term *functional response types* for species with a particular response to some environmental drivers. The traits of species that affect their abilities to perform under different environmental conditions are called response traits (*sensu* Elmqvist et al. 2003). Response diversity is the degree of variation in response traits found within a functional group of species (Norberg et al. 2001). On both evolutionary and ecological time scales, species sharing essential and limited resources experience competition and affect each others' growth and fitness. As a result, species with attributes that make them more efficient in the use of the resource will be more successful in competition and come to dominate (Tilman 1982). Thus, an autonomous selection process within the functional group of resource users shifts the composition of the whole group toward the most efficient combination of traits (Levin 1999). Obviously, such an adaptive process is related to the diversity of attributes of species or else there would be nothing to select among. In fact, it has been shown that the rate of the adaptive process is proportional to the variance in attributes (Fisher 1930; Norberg et al. 2001) and the relative difference between the performances of species with different attributes. In the recent abundant reports of ecological experiments on the role of biodiversity (generally the number of species) for ecosystem functioning (Schmid, Joshi, and Schläpfer 2003) a general phenomenon is the transitional effect under stable conditions when inferior species are outcompeted by more efficient ones (McGrady-Steed, Harris, and Morin 1997; Pacala and Tilman 1997; Norberg 2000). The emphasis here is on "under stable conditions." Since most experiments to date have tried to control environmental conditions as much as possible, this phenomenon has become known as a "sampling effect."

However, differences in efficiency between species depend on the existing conditions, i.e., their relative efficiencies may change as conditions change.

Under constant conditions, competition selects among the currently best-suited species and thus decreases response diversity locally (Walker, Kinzig, and Langridge 1999; Norberg et al. 2001; Norberg 2004) unless some spatial or resting/hibernation mechanism comes into play (Chesson 1994; Nystrŝm and Folke 2001; Elmqvist et al. 2003; Leibold and Norberg 2004). An important aspect of response traits is that traits for different types of disturbances may be correlated and the response capacity for one condition may change as a result of a change in another condition (Vinebrooke et al. 2004). It has been shown that response diversity is directly correlated to the capacity of a system to respond to changing conditions without a reduction in system functioning and that response diversity is determined by the nature of fluctuations in the environment (Norberg et al. 2001).

SOURCES OF DIVERSITY

Much focus in conservation efforts is put on sustaining a particular set of species under the assumption that this preserves biodiversity. Such efforts may be misguided, as they result in an artificially sustained species assemblage that may be maladapted to the continuously changing environment. Diversity is an aggregate measure and thus the components may change without changing the level of diversity. The level of diversity may also fluctuate in response to different drivers depending on the balance between the selective processes that reduce diversity and sources of diversity. In order to preserve the capacity of a community to change in response to changing conditions, the crucial focus for conservation efforts should be on sustaining the sources of diversity, rather than a particular set of species, which ensures that there is a continuous supply of species that can compete under current environmental conditions. A focus on such dynamic biodiversity means that the sources of diversity have to be identified.

The components of ecosystem diversity can most usefully be categorized at three scales: landscape (the diversity of communities within the landscape), species (the diversity of species within a community), and genetic (the diversity of genes—specifically alleles for the same gene—within a population of a species). We are not primarily interested in the patterns of diversity at these levels, but rather with mechanisms that promote diversity. Note that at the system level there are no obvious evolutionary mechanisms selecting for diversity, but physical and ecological processes can often result in more diverse communities. Landscape-level diversity consists of the variability in the land- or seascape, providing a multitude of different conditions for which different sets of attributes are best suited (see Cumming, Barnes, and Southworth, chapter 1, this volume, for a treatment of the subject of spatiotemporal asymmetries in landscapes). In a heterogeneous and changing landscape, dispersal, mobility, and ways of enduring unfavorable situations are an essential source of diversity in addition to complementarity between species as described in the previous section. Particularly in environments with high disturbance rates, competition between species may be reduced and locally suited ones may become extinct. Under such conditions, both species and structures that remain after the disturbance, ecological legacies, as well as those that can colonize from other locations, are crucial. During reorganization after disturbances the potential for input of a diversity of species directly determines the rate at which the system may self-organize toward a community that reflects the local conditions (Nyström and Folke 2001). Very high rates of immigration can, on the other hand, hinder local adaptation processes (Leibold and Norberg 2004). The concepts of spatial resilience and response diversity as outlined by Nyström and Folke (2001) and Elmqvist et al. (2003) capture many of the mechanisms involved. The sources of diversity at the landscape level have primarily to do with the spatiotemporal heterogeneity in environmental conditions and the capacity for dispersal/mobility or tolerance to unfavorable conditions over time, such as through resting stages or hibernation. Allopatric speciation is the landscape (biogeographic) effect on the evolutionary process that generates genetic diversity. These mechanisms are adaptations to spatiotemporal heterogeneity of the landscape.

The second level of diversity—species diversity—has received most attention from conservationists and is the most publicized form of change in diversity. Examples of diversity within a community of species, such as described by Walker, Kinzig, and Langridge (1999) for twenty-one grass species in an Australian rangeland ecosystem, provide clear evidence that species with quite different traits act together in a complementary way to enhance ecosystem function. Conversely, and importantly, the example also shows that the species in this community that are most similar in terms of their ecosystem function traits, but differ in terms of their responses to changing environmental conditions, confer resilience on ecosystem performance. Mechanisms that sustain diversity at the local community level, and that are not included in the landscape level effect discussed above, are those that enable coexistence in spatiotemporally invariant environments. These include multiple resources (Tilman 1982), niche separation along a resource aspect, i.e., complementarity (Hooper et al. 2005), and predator-mediated coexistence (e.g., Paine 1969). The evolutionary processes that cause genetic diversity at this scale are, for example, studied in the field of adaptive dynamics (Dieckmann and Doebeli 1999).

At the genetic level allelic diversity confers evolutionary potential in the long term. In the short term loss of alleles leads to reduced population fitness and low reproduction levels, as described for many different kinds of plant populations (cf. Harper 1977: chapter 24), leading to lowered population response diversity. As animal populations get smaller (through habitat loss and land clearing, for example), allelic diversity in the immune-defense system declines, and the population loses the capacity to withstand incursions of diseases. The variety of responses exhibited by different individuals within a population of a single species is directly due to the genetic diversity in that population. Reductions in genetic diversity can result from reduced population size or from reduced environmental variation—under constant environmental conditions alleles dominant under those conditions can lead to elimination of others alleles. Changes in genetic diversity are the least obvious of the changes in diversity, yet they are very significant in terms of effects on response diversity and therefore on the resilience of the populations concerned. Mechanisms that confer diversity at this level, in addition to the ones mentioned earlier, include sympatric speciation.

INSTITUTIONAL DIVERSITY AND GOVERNANCE

Institutions are used to regulate the actions of human agents in particular situations, such as fishermen doing their job. In its broadest sense the term *institution* refers to the regularized rules, norms, and strategies that agents use in making decisions in repeated situations at multiple levels of analysis (Crawford and Ostrom 1995). The concepts of a strategy, a norm, and a rule are linguistically related, as discussed below, but we will focus primarily on rules that structure repeated situations (or games) when we talk about institutional diversity. Governance involves crafting of rules in an effort to improve the incentives, behavior, and outcomes achieved in a situation over time—such as reducing the decline in fish populations by establishing protected areas.

A major confusion exists in the literature about institutions because rules operate and are made at multiple levels. When rules establish a protected area, for example, fishers must decide at an operational level where they will fish (Ostrom 1999). If they know the rules defining the protected area, consider these as legitimate rules that they wish to follow, and see that some monitoring effort is directed at anyone who might not honor the protected area, the fishers will go to some location other than the protected area. The fishers are acting at an operational level in their day-to-day decisions about where, when, and how much to fish. A fisher—or even all the fishers—may decide not to follow the rules if they are not considered legitimate and little effort is invested in monitoring and enforcement.

The rules that affect the incentives that actors face at an operational level are themselves made at a collective choice (or policy-making) level. Who is involved in a collective choice situation to craft the rules that affect incentives at an operational level varies from one fishery (or irrigation system, grazing area, or, more broadly, ecological system) to another. When the resource users have the authority to decide how a particular resource will be governed, they themselves may switch levels at specific times to determine whether they are satisfied or not with the rules that are currently in force. The resource users may themselves have the authority of policy makers for a particular resource. It is also frequent that a governmental unit operating at a local, regional, or national level is charged with the responsibility to engage in collective choice and to make operational level rules. The rules that affect the structure of a collective choice situation—e.g., who is involved, how they are chosen, which rules they can adopt and change, how much notice must they give to resource users, etc.—are themselves the result of decisions made at a constitutional level of analysis (Ostrom 1999). In this section of the chapter we will focus on institutional diversity, looking primarily at collective choice processes.

Since both natural and social systems are complex adaptive systems, it is usually impossible to predict particular outcomes that might arise from human interventions in the system. Most management situations, therefore, involve more or less deliberate processes of learning by experimentation. Adaptive management in particular tries to embrace processes of learning, evaluation, and change as an inherent solution of managing complex systems. Gunderson, Peterson, and Holling (chapter 8, this volume) state that "adaptive management usually is designed to be an iterative process that develops a social dialogue about a system. Modifying and creating ecological management institutions is difficult. Consequently, one of the main challenges of adaptive management is to develop new ways to enable social learning and experimentation."

Thus, a major component of successful adaptive management is a rich reservoir of options, i.e., institutional diversity. By institutional diversity we mean the range of operational and collective choice rules that have been tried out in diverse ecological systems. Just as biological diversity is important as a foundation for effective adaptations to occur, institutional diversity is also an important source of more effective adaptations (Ostrom 2005). Participants making rules have a higher likelihood of crafting effective rules when they perceive multiple options and can learn about how particular combinations of rules have helped or hindered other resource users solve challenging problems in similar ecological systems. All too often, however, scholars and policy makers make premature judgments that a particular type of rule—e.g., a transferable quota system—is the optimal rule to use in allocating rights to use a resource. Rigid beliefs about the existence of such panaceas—the one-size-fits-all approach—reduce the capacity of participants in a social system to learn about other rules that might fit a local situation better than the panacea frequently recommended (Evans 2004).

It is rarely, if ever, the case that one rule works everywhere in the same way. Such beliefs can cause major degradation of natural systems as evidence by many fishing resources that have been governed by MSY-determined catch limit regulations despite that fact that it has been impossible to construct a policy that ensures catch limits are actually followed (Jentoft, McCay, and Wilson 1998). Institutional diversity can be constructively used if there are effective methods by which evaluation and learning from trials can take place and if there is enabling legislation promoting this process (Berkes 2002). Below we elaborate on how institutional diversity can affect the process of adaptive management and help to generate more effective solutions.

ATTRIBUTES OF STRATEGIES, NORMS, AND RULES

In order to understand institutional diversity, one needs to understand how strategies, norms, and rules resemble one another as well as differ (Ostrom 2005: chapter 5). To do this, one needs a general classification scheme for the strategies, norms, and rules that individuals and groups adopt. An approach first developed by Crawford and Ostrom (1995) introduces an institutional grammar (ADICO) that may be used to classify these structured actions. According to this approach, five general variables help the analyst (as well as participants) sort out these closely related concepts. These are

A—Attributes of the person(s) involved,
D—whether a Deontic operator—a must, must not, or may—is involved,
I—the aIm of an action,
C—the Conditions involved, and
O—whether or not there is an Or else—a punishment.

A strategy involves only AIC and simply describes that individuals with particular attributes (such as pastoralists or fishers) try to achieve an aim (or purpose, such as putting animals out to pasture or catching fish) when some conditions are present (such as good weather). A norm—ADIC—involves adding a deontic to a basic strategy. A pastoralist may adopt a norm that he or she must take the animals to a different location every day or a fisher may adopt a norm that he or she must not fish in a region known as a spawning area. A rule involves authorized enforcement that adds an "or else." Pastoralists and farmers in a region may adopt a rule that pastoralists will bring their animals at a particular season when the farmers have only stubble left after their harvest and will graze their animals on the stubble for a defined period, providing fertilizer to the farmers, or else the fields will be closed to the pastoral group in the future. Thus a gradient in the level of sanctioning exists between a strategy (no sanction), a norm (internal shame or embarrassment accentuated by gossip), and a rule (a known sanction likely to be enforced by authorized individuals).

The components making up an operational or collective action situation may have different numeric, discrete, or logical values. For example, the target group of people to which a rule is applied, the amount one can harvest, the season when a rule is in effect, the severity of the sanction. These differences affect human behavior and how human actions affect ecological systems.

CONFLICTING AND COMPLEMENTARY INSTITUTIONS

The intended aim of an institution affects its designed function or effect. Conflicting or alternative institutions are characterized by overlapping aims

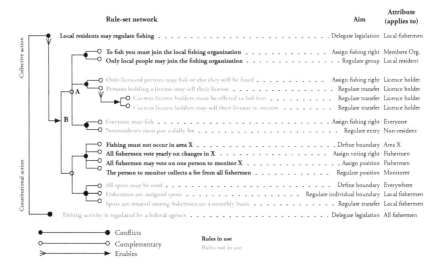

FIGURE 2.1. An example of a rule-set or policy network for regulating coastal fishing. The network shows not only the rules in use (black text) but also alternative rules currently not in use (gray text). The network illustrates how a diversity of alternative rules can be conflicting among rules with the same aim (filled circle connectors) and induce a choice, either a collective action decision on which rules to implement and enforce or simply which rules to ignore. Alternative conflicting rules or rule subsets are characterized by having the same aim and overlapping attributes, which lead to a conflict in choice. Complementary rules have separate aims. Some rules are needed to provide the aim for others, for example, the constitutional rights for local governance need to be in place in order for collective action to take place. In this example local regulation of fishing activity is allowed while the alternative, conflicting institution of federal regulation is not in use although it is a potential alternative. Note that a diversity of legislative organizations can lead to the simultaneous use of conflicting rules in different places at the same time and thus promote institutional diversity at larger spatial scales.

unless they are specified for a particular group of people, area, or time as determined by the attributes and conditions. If there is no conflict, the institutions might either be unrelated, complementary, or interdependent. The distinction is important since a conflict will lead to a choice situation about which rule to follow or to enforce and ultimately determine which rule comes to be used. Selective choices or processes are a fundamental aspect of adaptive systems in general and potential conflicts between institutions tell us where selection processes act. In this sense there is a strong parallel between the competition seen among species leading to succession within communities and the conflict between rules that lead to selective processes and succession within rule-sets. The outcome of these selective processes is a change in the relative importance of different rules, i.e., the degree of use.

To illustrate our reasoning we present in figure 2.1 an example of a rule-set network, which we will refer to in the following sections to elaborate greater detail about the role of institutional diversity for successful adaptive management. This network or rule-set has the general aim of regulating fishing effort, and particular solutions are shown as branching rule subsets. The three rule-subsets connected at point A all have the aim of assigning fishing rights (being members of a fishing organization, a license holder, or allowing open access), but they are in conflict if used at the same time. Nevertheless, they are (competing) alternatives, which could be tried out in successive experiments. In contrast, the two rule subsets connected at branch B are complementary in their aim. The aim of the upper rule sub-sets all regulate the assignment of the position of fishermen, while the lower rule subsets regulate the spatial extent in which fishing may occur. These aims do not conflict and they can be used simultaneously. Thus the degree of complementarity in the aim of institutions constrains institutional diversity in use, but not the diversity of institutions, which provide alternatives to the institutions in use. The rules that are not in use are in competition with the rules in use in the immediate term and at a particular place. However, they may provide the system with useful alternatives in the longer term or in other circumstances ore elsewhere at the same time.

INTERDEPENDENCIES OF INSTITUTIONS

All rules considered in figure 2.1 have the general aim of regulating fishing activity but differ in how they affect different behaviors and outcomes. The network illustrates how this general aim is broken down into more special-ized rules that may or may not interact. Direct interdependencies of rules may occur when one rule is needed for other rules to be in use, such as a constitutional-level rule that delegates policy-making authority to residents. Without this rule, all rules connected at point B are in conflict with federal fishing regulations.

Supporting rules may be created to regulate the effect of a previous rule. For example, if one rule allows participants to use nets, then there might be room for further rules that regulate mesh size, size of nets, and devices that decrease unwanted by-catches such as dolphins, thereby increasing the total diversity in rules. If there is a rule saying nets must not be used, then there is no need for these additional rules, but if the rule about no nets is conditional, i.e., dependent upon particular circumstances, then these additional rules may be dormant when those circumstances are extant but viable and used under other circumstances. The depth and the conditional branching of the rule structure implies finer adaptation and more detailed governance of activities. A deepening of the rule structure by supporting rules might be based on more extensive experiences learned from past experiments. Whether this leads to more sustainable resource use or more rigidity in the institutional structure and thus less adaptability is an open question.

Interdependent institutions are also inherently more sensitive to change because of the serialization of effects. For example, a constitutional rule that regulates the type of sanctioning local participants are allowed to use for collective action can have devastating effects on local governance if it is removed. One example comes from the Maine lobstermen who defend their group territories by cutting the traps of "intruders." However, in the past (no longer) state officials viewed trap cutting as destruction of property and thus sided with the fishermen who were actively destroying the territorial system. As a result, constraints on the escalation of fishing effort were removed and a rapid escalation followed.

Other types of interdependent institutions are those that regulate the spatial extent over which rules are to be applied, thus being a key rule for sustaining a diversity of local rules by preventing a one-size-fits-all approach, such as the Chilean Fishing and Aquaculture Law (Castilla and Fernandez 1998). Community involvement at a local level constitutes an effective tool by which fishers, scientists, and managers interact to improve the quality of the regulatory process and can create locally adapted institutional structures. There are many examples where nationalization of forests has had detrimental effects as local user groups have lost the power to regulate their use (Feeny 1988; Shepherd 1992; Gadgil and Iyer 1989; Grafton 2000; Jodha 1996).

INSTITUTIONAL DIVERSITY FOR RESPONDING TO CHANGING CONDITIONS

A great challenge for any policy maker is the fact that no natural system is stable or in "balance" (Holling 2001). Environmental change can affect the returns from the natural system or economic changes leading to rapid emerging markets for previously unexploited resources. More often than not, a different set of institutions will be needed to account for the changes in the feedback from the social-ecological system. Different strategies might be used under dry or rainy seasons (Holmgren and Scheffer 2001). If such changing conditions occur on a regular basis, alternative institutions can be crafted in advance. Kepe and Scoones (1999) show how institutions in use change depending on the state of a grassland system in South Africa, since the practitioners are well aware of the problems associated with each state and experience has lead to a suit of strategies that can be used under different conditions. Other examples include the "sleeping territorialities" of some Pacific Islands (Hviding 1989), the emergence of family-based hunting territorialities replacing a community-based one during the last two hundred years in the James Bay area (Berkes 1989), and the creation of special districts to replenish threatened groundwater basins in southern California (Blomquist 1992; Blomquist, Schlager, and Heikkila 2004).

Rarely occurring events pose an even more difficult challenge as institutions not in use will have a limited lifetime in the consciousness of the participants. Dale et al. (1998) stress the importance of institutional memory for dealing with infrequent disturbances. Knowledge systems and institutions may present a reservoir of alternative strategies that might be used more often under special circumstances (Berkes and Folke 2002; Agrawal 1999, 2005; Olsson, Folke, and Berkes 2004). Berkes (1998) retells the story of native Canadian Chisabi Cree Indians who experienced in 1984/85 a severe decline in caribou on their traditional hunting ground, possibly as a result of killing more than they could handle and not disposing of wastes in the years before. In a meeting some elders told of a similar decline some seventy years ago, possibly caused by the increased hunting efficiency of newly available repeating rifles. They advocated a responsible manner of hunting based on an underlying respect for the caribou. Wastes were cleaned up promptly. The caribou returned, and even expanded their former range. In this case, the norms of conduct were kept by the older generation and the social capital of the tribe included enough trust and respect for them to induce a change in behavior of the Cree hunting community.

CHANGING INSTITUTIONS

Selective processes occur when participants choose those rules to apply in a given situation, i.e., when there is competition among institutions.

As already discussed, rules have attributes, and different combinations of attributes may have different consequences. If the aim and the target group of two rules at the local level overlap, there is conflict, and the participants affected must choose which rule to follow or the participants involved in the collective action must decide which one to enforce. Thus rules having the same function—i.e., the same aim targeting the same group of people— cause a conflict situation, and a selection process must take place among the target group or the enforcing group of participants. As in the biological system, it is the relative expected performance of the rules that partly determines the outcome of the selection process. If the expected benefit of changing a rule outweighs the expected costs of this rule change (for details, see Ostrom 2005: chapter 8), a participant in the collective choice situation would have an incentive to vote for a change. However, there are also rules governing how collective action can take place. Collective choice rules may vary from reliance on the decision of a single leader, a small elite, to a majority or even consensus vote (Buchanan and Tullock 1962). Thus, the particular preferences of the participants and the consequences they expect from a rule change may differ, making the outcome of a vote dependent on the particular constellation of people (see also Anderies and Norberg, chapter 5, this volume, on information processes and social attractors).

SOURCES OF INSTITUTIONAL DIVERSITY

Local regulations may over time evolve to fit local conditions if legislative power has been assigned to local groups (Folke et al. 1998). Local rule-sets may over time reflect local conditions (Cumming, Barnes, and Southworth, chapter 1, this volume). However, due to historical constraints (path dependency) and local inventions, even local areas with similar conditions may evolve very different solutions. Information transfer between such areas then becomes a crucial source of institutional diversity. The ability to learn from other localities or situations may itself be subject to management, for example, by creating arenas for sharing of experiences or actively creating networks of people and groups, e.g., by bridging organizations (see Hahn et al., chapter 4, this volume). However, in simulation experiments it has been demonstrated that too much flow of information among local managers can reduce regional success in managing a resource sustainably by reducing the diversity of local solutions (Bodin and Norberg 2005). There probably exists a trade-off between sharing information by learning across scales and crafting locally suited solutions by trial and error. Both scientists as well as managers tend to gravitate toward the "best science" or "best practice" norms currently dominating the community and effectively reduce diversity by doing so. It has been shown, however, that increased diversity in the preference for different ideas may provide the trigger by which paradigm shifts can occur (Brock and Durlauf 1999).

Another source of institutional diversity is context transfer, which can occur by deliberate research and cognitive reasoning. We will exemplify this by the story of how the v-notch rule emerged in the Maine lobster fisheries. After WWII the state started rigorous enforcement of the prohibition on the sale of egged lobsters. Lobster buyers (dealers) then complained that many of the nonegged lobsters they purchased (legally) and put into live inventory became egged (and illegal) while in inventory (pounds). The state agreed to purchase these lobsters using license money and then returned the lobsters to the ocean. In order to stop unscrupulous lobstermen from catching these lobsters, washing the eggs off and reselling them, the lobsters were marked by the marine police with a v-notch in the tail, and v-notched lobsters were put in the same category as egged lobsters, i.e., illegal for sale. What happened next was interesting. Conservation-minded lobstermen realized that if they were to v-notch the egged lobsters they caught, other, unscrupulous, lobstermen would not be able to wash and sell these lobsters. In effect, the v-notch became a way for conservation-minded lobstermen to enforce conservation on other fishermen. The rule is very easy to enforce, and now most fishermen voluntarily—there is no legal requirement—notch an egged lobster when it is caught, and there is little or no problem with washing eggs. Some lobstermen deliberately notch undersized male lobsters for conservation purposes. Even though male, these are still illegal to sell.

Scientists have played a large role today in crafting institutions. While scientists can act as a source of new institutions, a widespread application of a regulation can also decrease diversity by outcompeting other alternatives. For example, the U.S. law on extended fisheries is a textbook application of the maximum sustainable yield concept (Clark 1976). Today details about how fishing activities are regulated have been modernized, but the underlying mental model for federal governance of fisheries remains largely based on the MSY concept despite its poor results in creating a sustainable fishery (Myers and Worm 2003). Current trends in promoting comanagement will likely have implications for how governments and nongovernment organizations will target their efforts in the future. Even co-management needs to be seen as one of the succession of experiments in the adaptive process.

New institutions may be introduced by people in local communities and put forth to the community council for a decision on whether to formalize them. New institutions can also take the form of voluntary imposed strategies among a few individuals. They could thus coexist as alternatives to existing rules in informal networks (see Hahn et al., chapter 4, this volume). During phases of change, when existing institutions fail to fulfill their aim, the reservoir of informal strategies and the experiences accumulated by people using them might be an important source of alternatives when decisions need to be made about new approaches to a particular management problem.

OPTIONS FOR MANAGING DIVERSITY IN SESS

In the preceding sections we have presented separate syntheses of how diversity can affect system performance in the ecological as well as institutional context. Many common mechanisms, however, determine why diversity emerges, what are the effects of diversity on system performance, and how selective processes act on diversity. In the following sections we will elaborate on how diversity can be managed using the insights from the above analysis.

MANAGING SPATIOTEMPORAL HETEROGENEITY

Stabilizing and homogenizing the environment, a frequent outcome of human intervention in ecosystems, provides a sense of certainty for people. Business plans and strategies, the coordination of market and nonmarket exchange, and technological and skill development are all easier to execute in a stable, predictable environment. In contrast, the diversity and ecological health of a species community is dependent upon a characteristic disturbance regime. Continuance of that regime allows "natural" levels of environmental fluctuation to continue and ensures that the community does not become dominated by a few species. When the conditions are stable over a long time, most complex adaptive systems inevitably lose diversity as units best suited to the prevailing, stable conditions become dominant at the expense of those less well suited to that particular set of conditions. One way to prevent such loss of diversity is to deliberately change the conditions. Inducing moderate disturbances can sustain adaptive capacity by maintaining diversity of units (Holling 2001). Competition is the main selective process that diminishes response diversity under periods of stability. If stable conditions are expected to prevail, diversity can thus be enhanced by deliberate reduction in competition between units. In ecology this is done by cropping, grazing, or other removal, thereby creating patches in which resources may be more abundant than in fully developed communities. During regrowth, species that cannot compete in the later stages of succession may temporarily become abundant and thereby increase the diversity of traits in the community.

In institutional settings, deliberate disturbances are achieved by changing the preferences on which decisions are based. Rotating national or local governments in democratic constitutions provides opportunities for changing individuals and thus the preferences they bring to the decision making process. By doing so, a reevaluation of old decisions may be achieved and new solutions may be established. Long-standing totalitarian governing systems may, on the other hand, become increasingly vulnerable (but, nevertheless, at least temporarily more efficient) as institutional solutions become more rigid if the worldview that forms the basis of decision making is kept streamlined and stable at all levels in the society.

Local diversity can be maintained by sustaining the process of local adaptation (or decision making), regional heterogeneity in environmental conditions, as well as mobility among local areas. To promote learning on a regional scale, information such as experiences or advice needs to flow between the local areas. However, too much flow can also disrupt the ability to learn by experimentation if a single solution becomes widespread regionally (Bodin and Norberg 2005). An analog problem is true in ecological metacommunities where immigration rates of species from outside that are too high may disrupt locally adapted communities (Leibold and Norberg 2004). A balance is needed between the effect of reducing diversity by selecting the locally best-suited species or rules and the influx of new ones from other areas. National fishing regulations enforced over large regions are, for example, in stark contrast to the diversity of institutional solutions that can be found in the local councils of Maine lobster fisheries (Acheson 1997) or the artisanal coastal fisheries in Chile enabled by national legislation (Castilla and Fernandez 1998).

ENABLING LOCAL COEXISTENCE

A basic tenet in ecology is that there cannot be more local consumer species than there are numbers of essential resources (Tilman 1982). However, due to resource specialization of consumers, even one essential resource can support several consumer species if it comes in many flavors (i.e., resource aspects). For example, water, an essential resource for plants, is distributed at different layers in the soil, which provides opportunity for many specializations of root length. In food webs diversity at one level may provide opportunity for sustaining diversity at higher levels in the food chain. Thus it seems reasonable to assume that maintaining diversity at the base level of the food chain will have effects for diversity of the entire food web.

In the institutional context, increasing the diversity of rules used in different governance systems may allow multiple ways of managing resources. Rules that tend to focus on only a single driver, for example, catch quantity rules in fishery, allow little room for adaptive response to changing circumstances (Acheson and Wilson 1996). Managing the diversity of *how* rules affect a certain behavior of appropriators rather than *how much* (i.e., catch quota) allows multiple simultaneous rules because the specific aims of such rules need not overlap. For example, rules that regulate no-take areas do not necessarily conflict with seasonal fishing regulations, although the general aim of both rules is to regulate fishing effort.

A way of increasing local diversity of rules is to increase the resource base of appropriators, for example, by providing a coastal area for general use rather than a set of defined species, in combination with enabling legislation that assigns collective actions right. If a fishing community is allowed to fish only for a few species, there will be fewer institutional solutions to harvesting of marine coastal resources than if the same community is allowed to harvest many different resources with different requirements for institutional arrangements. Even though the resource might be different, the capacity for learning and adaptation increases because of the human ability to draw and modify analogies from other circumstances.

Sometimes diversity is not as complementary in its effects, as described in the above sections. In both ecosystems and social systems, some species or rules respectively have greater consequences than others, i.e., are more critical for the performance of the system than others. The effects of a loss of such components have to do with how the network of interactions is shaped and may be hard to predict. Keystone species (Paine 1969) are ecological examples of such units, and human exploitation of these may have profound effects, as demonstrated by the effect of extinctions of otters on the kelp systems of the North American coasts (Steneck et al. 2002).

In social systems some rules have disproportionally large effects. Some rules are also more critical for the performance of a governance system than others. Dietz, Ostrom, and Stern (2003) review the empirical evidence regarding the performance of diverse forms of ownership of natural resources and find that all broad types of ownership (government, private, and common property) both succeed and fail to achieve sustainable use of natural resources. In the online supplement they illustrate their central findings that two types of rules appear to have preeminent importance for the success of any form of ownership. These rules carefully define the boundaries of a resource and who is authorized to use it and also define sanctions that are enforced and perceived to be legitimate by participants. Obviously, many types of boundary rules and enforced sanctions exist—just as there are many keystone species. Without enforced rules defining the boundary of a resource and the authorized users, however, it is almost impossible to gain agreement and compliance with other rules such as harvesting limits (see also Gibson, Williams, and Ostrom 2005; Hayes and Ostrom 2005).

BALANCING COSTS

There are many opportunities to sustain diversity, but all require some sacrifice or come with some cost because they all demand some form of restraint, some loss of short-term opportunities. Whether the cost of any particular rule or action is considered worthwhile depends upon the future benefits of increased adaptive capacity. In order for adaptive processes to be efficient, whether for species succession or transitions within rule-sets, there has to be some diversity for selective processes to operate on. At the same time, diversity, or the adaptive process itself, may be costly. In biological systems diversity can mean there are suboptimal species present that extract or hold up resources that otherwise more efficient species could use (Norberg et al. 2001). Agriculture solves such inefficiencies by deliberately eliminating diversity and only planting the best crop—best at least in the short run. Suboptimal species may, of course, become the most optimal species if environmental conditions change, but the point is that the community usually has a cost in terms of the resources extracted by "passenger species" that decrease instantaneous efficiency but may increase long-term efficiency in a variable environment (Walker, Kinzig, and Langridge 1999; Norberg et al. 2001).

In an institutional setting the costs of diversity arise in at least two ways. First, there is a cognitive cost in predicting and evaluating expected outcomes of different rules. Since any rule might have a large number of different outcomes (depending on the configuration of other rules and ecological conditions present), this task can be too complex for any human being, let alone a group, to analyze. Instead, simple rules of thumb or heuristics derived from experience or analogy with other circumstances can be applied that only crudely approximate the expected consequences of a rule but are manageable (Wilson 2002). Second, there is the collective cost of actually changing the rules in use, such as the upfront time and effort spent on devising and agreeing upon new rules, the short-term costs of adopting new appropriation strategies and the long-term costs of monitoring and enforcing the use of the new rule (Ostrom 2005: chapter 8). In the latter case, the cost is associated with the process of change, rather than with sustaining a diversity of options.

Managing diversity in social-ecological systems is a daunting task for several reasons. First, the effect of increased or sustained diversity is not easily predicted and depends on the particular attributes of the units involved and how they interact, as we have outlined above. It is seldom possible to affect only one aspect, such as the range of responses to environmental drivers in species, without also affecting other aspects that might cause secondary, unpredictable effects. Second, the argument for diversity is continuously in conflict with a worldview that conceives of natural systems as simple, homogenous resources that are predictable and subject to optimization. From this perspective, maintaining natural diversity is costly to human well-being. Its redundancies and plethora of things that aren't immediately useful are not seen as an insurance policy for the future. Third, in order to sustain diversity in social-ecological systems, one needs to address the sources of diversity, which are often difficult to identify and may involve consideration of multiple spatial scales. This means that a focus on diversity for management has to be targeted toward sustained generation of alternatives rather than maintenance of particular sets of species or rules. This dynamic view of diversity challenges traditional conservation efforts that typically focus on preservation of endangered species and prevention of invasive species.

Social-ecological systems are complex adaptive systems. Although one cannot predict exactly what outcome the introduction or removal of a particular species or rule might have, there is a general understanding of how diversity increases a system's ability to cope with change, reduces its sensitivity to loss of components, and creates a more satisfying experience for human well-being. We argue that two general methods for sustaining diversity in social-ecological systems can act as general guidelines for managers: 1. Promoting local adaptations, such as local crop varieties and the associated knowledge of how to use them or enabling co-management, which promotes locally crafted solutions for managing resources, creates a regional diversity of species, experiences, and knowledge that can be transferred and shared among local areas. Too much sharing, however, such as nationally enforced "blueprint solutions" or "management fashions" spread by eager NGOs, may reduce regional diversity. 2. Enabling the diversity of local governance or decision units in order to minimize the dominance of single solutions such as might occur with multiple sustainable uses of coastal resources or forest products. This could involve stakeholder identification and participation (see Lebel and Bennett, chapter 7, this volume) in the decision process and defined rights of use and legislation to the local appropriators. Adaptive management (see Gunderson, Peterson, and Holling, chapter 8, this volume) provides a method by which solutions are tested and exchanged over time. Another method is to use multiple institutional solutions for one aim at any one time, such as having both gear as well as seasonal and spatial restrictions on the fishing effort. All these methods avoid the trap of letting one solution dominate and provide a richer experience and knowledge base, enhancing the chances of improving the capacity of society to make better decisions. Thus diversity in general increases the capacity of social-ecological systems to tolerate disturbance, learn, and change. This capacity will be one of the most crucial assets of societies in the coming times of rapid global change

REFERENCES

Acheson, J. 1997. "The Politics of Managing the Maine Lobster Industry: 1860 to the Present." *Human Ecology* 25(1): 3–27.

Acheson, J., and J. Wilson. 1996. "Order Out of Chaos: The Case for Parametric Fisheries Management." *American Anthropologist* 98(3): 579–594.

Agrawal, A. 1999. *Greener Pastures: Politics, Markets, and Community Among a Migrant Pastoral People*. Durham: Duke University Press.

Agrawal, A. 2005. *Environmentality: Technologies of Government and the Making of Subjects*. Durham: Duke University Press.

Anderies, J. M., M. Janssen, and E. Ostrom. 2004. "A Framework to Analyze the Robustness of Social-Ecological Systems from an Institutional Perspective." *Ecology and Society* 9(1): 18.

Arthur, W. B., S. N. Durlauf, and D. A. Lane, eds. 1997. *The Economy as an Evolving Complex System II*. Boston: Addison-Wesley.

Bai, Y. F., X. G. Han, J. G. Wu, Z. Z. Chen, and L. H. Li. 2004. "Ecosystem Stability and Compensatory Effects in the Inner Mongolia Grassland." *Nature* 431:181–184.

Berkes, F., ed. 1989. *Common Property Resources: Ecology and Community-Based Sustainable Development*. London: Belhaven.

Berkes, F. 1998. "Indigenous Knowledge and Resource Management Systems in the Canadian Subarctic." In F. Berkes and C. Folke, eds., *Linking Social and Ecological Systems: Management Practices and Social Mechanisms for Building Resilience*, pp. 98–128. Cambridge: Cambridge University Press.

Berkes, F. 2002. "Cross-scale Institutional Linkages: Perspectives from the Bottom Up." In E. Ostrom, T. Dietz, N. Dolsak, P. C. Stern, S. Stonich, and E. U. Weber, eds., *The Drama of the Commons*, pp. 293–321. Washington, DC: National Academy Press.

Berkes, F., and C. Folke. 2002. "Back to the Future: Ecosystem Dynamics and Local Knowledge." In L. H. Gunderson and C. S. Holling, ed., *Panarchy: Under-*

standing Transformations in Human and Natural Systems, pp. 121–146. Washington, DC: Island.

Blomquist, W. 1992. *Dividing the Waters: Governing Groundwater in Southern California*. San Francisco: ICS.

Blomquist, W., E. Schlager, and T. Heikkila. 2004. *Common Waters, Diverging Streams: Linking Institutions and Water Management in Arizona, California, and Colorado*. Washington, DC: Resources for the Future.

Bodin, O., and J. Norberg. 2005. "Information Network Topologies for Enhanced Local Adaptive Management." *Environmental Management* 35(2): 175–193.

Brock, W., and S. N. Durlauf. 1999. A formal model of theory choice in science. Economic Theory 14(1): 113–130.

Brown, J. H., T. G. Whitham, S. K. Morgan Ernest, C. A. Gehring. 2001. "Complex Species Interactions and the Dynamics of Ecological Systems: Long-term Experiments." *Science* 293:643–650.

Buchanan, J. M., and G. Tullock. 1962. *The Calculus of Consent*. Ann Arbor: University of Michigan Press.

Carpenter, S. R., B. H. Walker, J. M. Anderies, and N. Abel. 2001. "From Metaphor to Measurement: Resilience of What to What?" *Ecosystems* 4:765–781.

Castilla, J. C., and M. Fernandez. 1998. "Small-scale Benthic Fisheries in Chile: On Co-management and Sustainable Use of Benthic Invertebrates." *Ecological Applications* 8:124–132.

Chase, J. M, and M. A. Leibold. 2003. *Ecological Niches: Linking Classical and Contemporary Approaches*. Chicago: University of Chicago Press.

Chesson, P. 1994. "Multispecies Competition in Variable Environments." *Theoretical Population Biology* 45:227–276.

Chesson, P. 2000. "Mechanisms of Maintenance of Species Diversity." *Annual Review of Ecology and Systematics* 31:343–366.

Clark, C. W. 1976. *Mathematical Bioeconomics: The Optimal Management of Renewable Resources*. New York: Wiley.

Crawford, S. E. S., and E. Ostrom. 1995. "A Grammar of Institutions." *American Political Science Review* 89(3): 582–600.

Daily, G. C. 1997. "Introduction: What Are Ecosystem Services?" In G. C. Daily, ed., *Nature's Services: Societal Dependence on Natural Ecosystems*, pp. 1–10. Washington, DC: Island.

Dale, V. H., A. E. Lugo, J. A. MacMahon, and S. T. A. Pickett. 1998. "Ecosystem Management in the Context of Large, Infrequent Disturbances." *Ecosystems* 1:546–557.

Diaz, S., and Cabido, M. 2001. "Vive la Difference: Plant Functional Diversity Matters to Ecosystem Functioning." *Trends in Ecology and Evolution* 16:646–655.

Dieckmann, U., and M. Doebeli. 1999. "On the Origin of Species by Sympatric Speciation." *Nature* 400:354–357.

Dietz, T., E. Ostrom, and P. Stern. 2003. "The Struggle to Govern the Commons." *Science* 302(5652): 1907–1912.

Dunne, J.A., R.J. Williams, and N.D. Martinez. 2002. "Network Structure and Biodiversity Loss in Food Webs: Robustness Increases with Connectance." *Ecology Letters* 5:558–567.

Edelman, G. M., and J. A. Gally. 2001. "Degeneracy and Complexity in Biological Systems." *PNAS* 98:3763–1376.

Elmqvist, T., C. Folke, M. Nyström, G. Peterson, J. Bengtsson, B. Walker, and J. Norberg. 2003. "Response Diversity, Ecosystem Change, and Resilience." *Frontiers in Ecology and the Environment* 1(9): 488–494.

Emmerson, M.C., M. Solan, C. Emes, D.M. Paterson, and D. Raffaelli. 2001. "Consistent Patterns and the Idiosyncratic Effects of Biodiversity in Marine Ecosystems." *Nature* 411(6833): 73–77.

Evans, P. 2004. "Development as Institutional Change: The Pitfalls of Monocropping and Potentials of Deliberation." *Studies in Comparative International Development* 38(4): 30–53.

FAO (Food and Agriculture Organization of the United Nations).1993. *Harvesting Nature's Diversity.* Rome: FAO.

Fardon, R. 1995. *Counterworks: Managing the Diversity of Knowledge.* London: Routledge.

Feeny, D. 1988. "Agricultural Expansion and Forest Depletion in Thailand, 1900–1975." In J.F. Richards and R.P. Tucker, eds., *World Deforestation in the Twentieth Century*, pp. 112–143. Durham: Duke University Press.

Fisher, R. A. 1930. *The Genetical Theory of Natural Selection.* Oxford: Clarendon.

Folke, C., J. Colding, and F. Berkes. 2003. "Synthesis: Building Resilience and Adaptive Capacity in Social-Ecological Systems." In F. Berkes, J. Colding, and C. Folke, eds., *Navigating Social-Ecological Systems: Building Resilience for Complexity and Change*, pp. 352–387. Cambridge: Cambridge University Press.

Folke, C., L. Pritchard Jr., F. Berkes, J. Colding, and U. Svedin. 1998. "The Problem of Fit Between Ecosystems and Institutions." IHDP Working Paper no 2. http://www.ihdp.uni-bonn.de/html/publications/workingpaper/wp02m.htm.

Folke, C., S. Carpenter, B. Walker, M. Scheffer, T. Elmqvist, L. Gunderson, and C.S. Holling. 2004. "Regime Shifts, Resilience, and Biodiversity in Ecosystem Management." *Annual Review of Ecology Evolution and Systematics* 35: 557–81.

Gadgil, M., and P. Iyer. 1989. "On the Diversification of Common-Property Resource Use by Indian Society." In F. Berkes, ed., *Common Property Resources: Ecology and Community-Based Sustainable Development*, pp. 240–272. London: Belhaven.

Gibson, C.C., J. T. Williams, and E. Ostrom. 2005. "Local Enforcement and Better Forests." *World Development* 33(2):273–284.

Grafton, R. Q. 2000. "Governance of the Commons: A Role for the State." *Land Economics* 76(4):504–517.

Grime, J. P. 1997. "Biodiversity and Ecosystem Functioning: The Debate Deepens." *Science* 277:1260–1261.

Grime, J. P., K. Thompson, R. Hunt, J. G. Hodgson, J. H. C. Cornelissen, I. H. Rorison, G. A. F. Hendry, T. W. Ashenden, A. P. Askew, S. R. Band, R. E. Booth et al. 1997. "Integrated Screening Validates Primary Axes of Specialisation in Plants." *Oikos* 79:259–281.

Gunderson, L., and C. Holling. 2002. *Panarchy: Understanding Transformations in Human and Natural Systems.* Washington, DC: Island.

Hairston, N. G., F. E. Smith, and L. B. Slobodkin. 1960. "Community Structure, Population Control, and Competition." *American Naturalist* 94(87): 421–425.

Harper, J. L. 1977. *Population Biology of Plants.* London: Academic.

Hayes, T. M., and E. Ostrom. 2005. "Conserving the World's Forests: Are Protected Areas the Only Way?" *Indiana Law Review* 38(3): 595–617.

Heemsbergen, D. A., M. P. Berg, M. Loreau, J. R. van Hal, J. H. Faber, and H. A. Verhoef. 2004. "Biodiversity Effects on Soil Processes Explained by Interspecific Functional Dissimilarity." *Science* 306:1019–1020.

Hodgson, G. M. 2002. "Darwinism in Economics: From Analogy to Ontology." *Journal of Evolutionary Economics* 12(3): 259–281.

Holland, J. 1995. *Hidden Order: How Adaptation Builds Complexity.* Boston: Addison-Wesley.

Holling, C. S. 1959a. The Components of Predation as Revealed by a Study of Small Mammal Predation of the European Pine Sawfly." *Canadian Entomologist* 91:293–320.

Holling, C. S. 1959b. "Some Characteristics of Simple Types of Predation and Parasitism." *Canadian Entomologist* 91:385–398.

Holling, C. S. 1973. "Resilience and Stability of Ecology Systems." *Annual Review of Ecology and Systematics* 4:1–23.

Holling, C. S. 1992. "Cross-scale Morphology, Geometry and Dynamics of Ecosystems." *Ecological Monographs* 62(4): 477–502.

Holling, C. S. 2001. "Understanding the Complexity of Economic, Ecological, and Social Systems." *Ecosystems* 4:390–405.

Holmgren, M., and M. Scheffer. 2001. "El Niño as a Window of Opportunity for the Restoration of Degraded Ecosystems." *Ecosystems* 4:151–159.

Holmgren, M., M. Scheffer, and M. A. Huston. 1997. "The Interplay of Facilitation and Competition in Plant Communities." *Ecology* 78(7): 1966–1975.

Hooper, D. U. 1998. "The Role of Complementarity and Competition in Ecosystem Response to Variation in Plant Diversity." *Ecology* 79:704–719.

Hooper, D. U., F. S. Chapin III, J. J. Ewel, A. Hector, P. Inchausti, S. Lavorel, J. H. Lawton, D. M. Lodge, M. Loreau, S. Naeem, S. Schmid, H. Setälä,

A. J. Symstad, J. Vandermeer, A. Wardle. 2005. "ESA Report: Effects of Biodiversity on Ecosystem Functioning, a Consensus of Current Knowledge. *Ecological Monographs* 75(1):3–35.

Hubbell, S. P. 2001. *The Unified Neutral Theory of Biodiversity and Biogeography.* Princeton: Princeton University Press.

Hughes, T. P., A. H. Baird, D. R. Bellwood, M. Card, S. R. Connolly, C. Folke, R. Grosberg, O. Hoegh-Guldberg, J. B. C. Jackson, J. Kleypas, J. M. Lough, P. Marshall, M. Nyström, S. R. Palumbi, J. M. Pandolfi, B. Rosen. 2003. "Climate Change, Human Impacts, and the Resilience of Coral Reefs." *Science* 301:929–933.

Hulot, F. D., G. Lacroix, F. Lescher-MoutouŽ, and M Loreau,. 2000. "Functional diversity Governs Ecosystem Response to Nutrient Enrichment." *Nature* 405:340–344.

Hviding, E. 1989. *All Things in Our Sea: The Dynamics of Customary Marine Tenure, Marovo Lagoon, Solomon Islands.* NRI Special Publication no. 13. Boroko: National Research Institute.

Janssen, M. 1998. "Use of Complex Adaptive Systems for Modeling Global Change." *Ecosystems* 1:457–463.

Jentoft, S., B. J. McCay, and D. C. Wilson. 1998. "Social Theory and Fisheries Comanagement." *Marine Policy* 22(4–5): 423–436.

Jodha, N. S. 1996. "Property Rights and Development." In S. Hanna, C. Folke, and K.-G. MŠler, eds., *Rights to Nature*, pp. 205–222. Washington, DC: Island.

Jones, C. G., J. H. Lawton, and M. Shachak. 1994. "Organisms as Ecosystem Engineers." *Oikos* 69:373–386.

Kepe, T., and I. Scoones. 1999. "Creating Grasslands: Social Institutions and Environmental Change in Mkambati Area, South Africa." *Human Ecology* 27(1): 29–53.

Koestler, A. 1967. *The Ghost in the Machine.* London: Pan.

Langton, C. G. 1995. *Artificial Life: An Overview.* Cambridge: MIT Press.

Leibold, M. A. 1995. The Niche Concept Revisited: Mechanistic Models and Community Context." *Ecology* 76:1371–1382.

Leibold, M. A., and J. Norberg. 2004. "Biodiversity in Metacommunities: Plankton as Complex Adaptive Systems?" *American Society of Limnology and Oceanography* 49(4) (part 2): 1278–1289.

Levin, S. A. 1998. "Ecosystems and the Biosphere as Complex Adaptive Systems." *Ecosystems* 1(5): 431–436.

Levin, S. A. 1999. *Fragile Dominion: Complexity and the Commons.* Reading, MA: Perseus.

Loreau, M., and A. Hector. 2001. "Partitioning Selection and Complementarity in Biodiversity Experiments." *Nature* 412:72–76.

Loreau, M., S. Naeem, P. Inchausti, J. Bengtsson, J. P. Grime, A. Hector, D. U. Hooper, M. A. Huston, D. Raffaelli, B. Schmid, D. Tilman, and D. A. Wardle.

2001. "Biodiversity and Ecosystem Functioning: Current Knowledge and Future Challenges." *Science* 294:804–808.

Low, B., E. Ostrom, C. Simon, and J. Wilson. 2003. "Redundancy and Diversity: Do They Influence Optimal Management?" In F. Berkes, J. Colding, and C. Folke, eds., *Navigating Social-Ecological Systems: Building Resilience for Complexity and Change*, pp. 83–114. Cambridge: Cambridge University Press.

MacArthur, R. H., and R. Levins. 1967. The Limiting Similarity, Convergence, and Divergence of Coexisting Species." *American Naturalist* 101:377–385.

McGrady-Steed, J., P. M. Harris, and P. J. Morin. 1997. "Biodiversity Regulates Ecosystem Predictability." *Nature* 390:162–165.

Myers, R. A., and B. Worm. 2003. "Rapid Worldwide Depletion of Predatory Fish Communities." *Nature* 425:280–283.

Naeem, S. and J. P. Wright. 2003. "Disentangling Biodiversity Effects on Ecosystem Functioning: Deriving Solutions to a Seemingly Insurmountable Problem." *Ecology Letters* 6:569–579.

Norberg J. 2000. Resource-Niche Complementarity and Autotrophic Compensation Determines Ecosystem-Level Responses to Increased Cladoceran Species Richness." *Oecologia* 122(2): 264–272.

Norberg, J. 2004. Biodiversity and Ecosystem Functioning: A Complex Adaptive Systems Approach." *American Society of Limnology and Oceanography* 49(4) (part 2): 1269–1277.

Norberg, J., D. P. Swaney, J. Dushoff, J. Lin, R. Casagrandi, and S. A. Levin. 2001. "Phenotypic Diversity and Ecosystem Functioning in Changing Environments: A Theoretical Framework." *PNAS* 98(20): 11376–11381.

Nyström, M., and C. Folke. 2001. Spatial Resilience of Coral Reefs." *Ecosystems* 4:406–417.

Olsson, P., C. Folke, and F. Berkes. 2004. "Adaptive Co-management for Building Resilience in Social-Ecological Systems." *Environmental Management* 34(1): 75–90.

Ostrom, E. 1998. "A Behavioral Approach to the Rational Choice Theory of Collective Action." *American Political Science Review* 92(1): 1–22.

Ostrom, E. 1999. "Institutional Rational Choice: An Assessment of the Institutional Analysis and Development Framework." In P. A. Sabatier, ed., *Theories of the Policy Process*, pp. 35–71. Boulder: Westview.

Ostrom, E. 2005. *Understanding Institutional Diversity*. Princeton: Princeton University Press.

Ostrom, E., R. Gardner, and J. Walker. 1994. Rules, *Games, and Common-Pool Resources*. Ann Arbor: University of Michigan Press.

Pacala, S. W., and S. A. Levin. 1997. "Biologically Generated Spatial Pattern and the Coexistence of Competing Species." In D. Tilman and P. Kareiva, eds., *Spatial Ecology: The Role of Space in Population Dynamics and Interspecific Interactions*, pp. 204–232. Princeton: Princeton University Press.

Pacala, S.W., and D. Tilman. 1997. "The Transition from Sampling to Complementarity." In D. Tilman and P. Kareiva, eds., *Spatial Ecology: The Role of Space in Population Dynamics and Interspecific Interactions*, pp. 151–166. Princeton: Princeton University Press.

Pace, M.L., J.J. Cole, S.R. Carpenter, and J.F. Kitchell. 1999. "Trophic Cascades Revealed in Diverse Ecosystems." *Trends in Ecology and Evolution* 14(12): 483–488.

Paine, R.T. 1969. "A Note on Trophic Complexity and Species Diversity." *American Naturalist* 100:91–93.

Petchey, O.L., and K.J. Gaston. 2002. "Functional Diversity (FD), Species Richness, and Community Composition." *Ecology Letters* 5:402–411.

Petchey, O.L., A. Hector, and J. Gaston. 2004. "How Do Different Measures of Functional Diversity Perform?" *Ecology* 85(3): 847–857.

Scheffer, M., and S.R. Carpenter. 2003. "Catastrophic Regime Shifts in Ecosystems: Linking Theory to Observation." *Trends in Ecology and Evolution* 18(12): 648–656.

Schmalensee, R. 1972. "Option Demand and Consumer's Surplus: Valuing Price Changes under Uncertainty." *American Economic Review* 62(5): 813–824.

Schmid, B., J. Joshi, and F. Schläpfer. 2003. "Empirical Evidence for Biodiversity-Ecosystem Functioning Relationships." In A. Kinzig, S.W. Pacala, and D. Tilman, eds., *The Functional Consequences of Biodiversity*, pp. 120–151. Princeton: Princeton University Press.

Shepherd, Gill. 1992. *Managing Africa's Tropical Dry Forests: A Review of Indigenous Methods*. London: Overseas Development Institute.

Steneck, R.S. 2001. "Functional Groups." In S.A. Levin, ed., *Encyclopedia of Biodiversity*, pp. 121–139. Boston: Academic.

Steneck, R.S., M.H. Graham, B.J. Bourque, D. Corbett, J.M. Erlandson, J.A. Estes, and M.J. Tegner. 2002. "Kelp Forest Ecosystems: Biodiversity, Stability, Resilience, and Future." *Environmental Conservation* 29(4): 436–459.

Tilman, D. 1982. *Resource Competition and Community Structure*. Princeton: Princeton University Press.

Tilman, D., C.L. Lehman, and K.T. Thomson. 1997. "Plant Diversity and Ecosystem Productivity: Theoretical Considerations." *PNAS* 94:1857–1861.

Vinebrooke, R.D., K.L. Cottingham, J. Norberg, M. Scheffer, S.I. Dodson, S.C. Maberly, and U. Sommer. 2004. "Impacts of Multiple Stressors on Biodiversity and Ecosystem Functioning: The Role of Species Co-tolerance." *Oikos* 104(3): 451–457.

Walker, B., A. Kinzig, and J. Langridge. 1999. "Plant Attribute Diversity, Resilience, and Ecosystem Function: The Nature and Significance of Dominant and Minor Species." *Ecosystems* 2(2): 95–113.

Walker, B., and J. Meyers. 2004. "Thresholds in Ecological and Social-Ecological Systems: A Developing Database." *Ecology and Society* 9(2): 3.

Walker, B., C. S. Holling, S. R. Carpenter, and A. Kinzig. 2004. "Resilience, Adaptability and Transformability in Social-Ecological Systems." *Ecology and Society* 9(2): 5. http://www.ecologyandsociety.org/vol9/iss2/art5.

Wilson, James. 2002. "Scientific Uncertainty, Complex Systems, and the Design of Common Pool Institutions." In E. Ostrom, T. Dietz, N. Dolsak, P.C. Stern, S. Stonich, and E. U. Weber, eds., *The Drama of the Commons*, pp. 293–321. Washington, DC: National Academy Press.

Worm, B., and J. E. Duffy. 2003. "Biodiversity, Productivity, and Stability in Real Food Webs." *Trends in Ecology and Evolution* 18(12): 628–632.

INTRODUCTION TO PART 2
NETWORKS

NETWORKS ARE INTEGRAL to our daily lives. They include such diverse examples as blood and organ systems, road and rail networks, rivers, food webs, telephone networks, local book clubs, and the complex global network of the Internet. All networks can be described as sets of nodes linked by lines or "edges." Nodes typically represent entities such as people, animals, places, or organs, while edges represent connections or interactions between nodes. The study of networks is a transdisciplinary field with influences from a variety of disciplines that include social science, engineering, physics, and biology.

Network theory deals with the general properties of networks. It has focused on 1. network structure and topology, which describe how nodes are connected; 2. emergent properties of networks, such as scale-free networks and degree link distributions; 3. the dynamics of network assembly, such as preferential attachments to highly connected nodes or disassembly processes such as targeted removals, e.g., by sabotage of transport networks; and 4. the effects of network configuration on some measures of performance of the network, such as the transfer efficiency of the network or its robustness to node removal. A fundamental aspect of complex adaptive systems is that localized interactions (i.e., interactions that are constrained in space and/or time) can create selective processes. These processes act on a range of the characteristics of network components. While local interactions can lead to local adaptation, they may also constrain the ability of network components (and entire networks) to reach global optima. Consequently, understanding the dynamics of networks of interactions is integral to understanding the dynamics of complex systems.

Networks often exist to achieve particular objectives or functions. The topology of the network influences its functionality through such details as the strength and permanence of links and nodes, how they are formed, how they change in response to the dynamics of other nodes or external factors, and networkwide properties such as the proportions of parallel and serial connections. For example, the habitat of many species consists of suitable patches that are embedded in a less hospitable matrix. The degree of connectivity between patches can determine whether a species can survive in such a landscape. Furthermore, the potential of the landscape itself as a living space for a species or metapopulation may depend on a few important patches that connect otherwise isolated clusters of patches. The presence of multiple connections in such a patch network relates directly to the sensitivity of the whole network to habitat loss or regeneration.

Colleen Webb and Örjan Bodin discuss some important emergent properties of networks and explain what they mean for management and sustainability. Emergent properties are phenomena that are greater than the sum of the parts. In networks they are manifested when nodes, such as people or organisms, are connected to each other through the exchange of matter or information. The topology of a network may have emergent implications for the performance of the network as a whole. For example, in "small world" networks, each node is linked to each other node by a relatively small number of connections. For human societies this number is typically around 6. In "scale-free" networks any small part of the network will look similar in shape to a larger part of the same network. The influence of perturbations on networks depends partially on the nature of nodes and links; node diversity and node redundancy, respectively the degree of different and shared properties among nodes in the network, are important system attributes, as discussed earlier by Norberg et al., chapter 2, this volume. In scale-free systems most nodes have few connections while others have disproportionally many. Highly connected nodes are very efficient in connecting remote parts of networks, but they also increase the sensitivity of a network to perturbations. These nodes are critical components that can have large effects on the rest of the network if they are removed or cease to function and thus have large implications for the resilience of the network.

Network structure can also affect the resilience of a system in other ways. Control of flow among nodes and the modular structure of a network may reduce the impact of perturbations by creating barriers to their spread. In a system consisting of completely distinct, isolated clusters, a perturbation cannot spread beyond the initial site of initiation. However, resilience is also determined by the ability of nodes and links to adapt and learn, both of which may become strongly constrained by modular structures. Fulfillment of these dual requirements suggests that network resilience is enhanced by a balance between a completely modular structure and a globally connected one. The chapter by Webb and Bodin goes into some depth on explaining how to analyze and understand the role of modularity and how to identify nodes that may be particularly important for targeted management in order to sustain the functioning of the network.

Social networks are central to sustainable governance. For example, trust between people and people's expectations regarding offers of cooperation are central for governance that relies on local initiatives for natural resource management. The chapter by Hahn et al. provides some general ideas about how social networks affect successful governance and presents a concrete example of a success story. Governance in social-ecological systems involves a large number of key players, with different skills, personalities, and preferences, as well as the representation of different stakeholder groups. Interactions among these people and groups may be crucial for learning to successfully manage a resource, but can also lead to conflict, causing polarization of groups and deadlocks with devastating results for the overall effectiveness of management efforts.

In a seminal paper published in 1973, Mark Granovetter argued that weak ties (e.g., the bridges between different stakeholder groups) may be the most valuable asset for generating new knowledge and identifying new opportunities: "those to whom we are weakly tied are more likely to move in circles different from our own and will thus have access to information different from that which we receive." Thus, networks can be very useful for knowledge generation, mobilizing people, identifying common interests, and starting projects. The formation of a network, however, is no guarantee of resilience or effective management; problem solving within a network depends on the content of exchanges and the kinds of action that they initiate. The chapter by Hahn et al. illustrates some interesting aspects of navigating a social system, identifying important groups of people, and bringing people together with the goals of sense making, conflict resolution, and building a common vision. One lesson learnt from Kristianstads Vattenrike, a Man and the Biosphere Reserve in southern Sweden, is that a clear vision and direction for ecosystem-based development can coexist with an exceptionally flexible project organization.

3

A NETWORK PERSPECTIVE ON MODULARITY AND CONTROL OF FLOW IN ROBUST SYSTEMS

Colleen Webb and Örjan Bodin

ECOLOGICAL AND SOCIAL systems are not the only complex adaptive systems (CAS) of interest to scientists. Across disciplines, ranging from computer science and engineering to immunology, the study of CAS emerges with important relevance to questions of robustness. Like resilience, robustness has multiple, often conflicting definitions in the literature. Despite difficulties with definition, the concept of robustness generally focuses on the persistence of particular features of a system in the face of a variety of perturbations (Jen 2005). This idea is clearly related to the definition of resilience used in this book (Holling 1973). Within this chapter, we choose to deal with robustness rather than directly with resilience because many network approaches to system response to disturbance are not literally in the context of resilience (*sensu* Holling 1973).

The CAS theory and empirical results suggest four general mechanisms of robustness: diversity of the system's parts, redundancy of these parts, control of flows within the system, and modular structure (Webb and Levin 2005). Ecosystems (Levin 1998) and various social systems fulfill the criteria of complex adaptive systems, and hence we can hypothesize that these four mechanisms of robustness in CAS may also be important in ecological and social systems. Previous chapters (chapters 1 and 2) have dealt with the importance of diversity and redundancy in maintaining system function following disturbance. Diversity and redundancy facilitate the replacement of components lost to disturbance with other components that can perform the same function (Ehrlich and Walker 1998). In contrast to diversity and redundancy, control of flow and modular structure directly maintain system function by reducing the impact of disturbance in the first place.

Control of flow and modular structure reduce the impact of disturbance by creating barriers to its spread. In a system consisting of completely distinct, isolated clusters, the spread of disturbance cannot occur beyond the initial site of disturbance. However, in both social and ecological systems, individuals within these clusters depend on the flow of information and resources from throughout the entire system. Despite the apparent advantage to a modular structure, fulfillment of these dual requirements suggests a balance between a completely modular structure and a globally connected one. If modular structure (i.e., modularity) and control of flow are mechanisms of robustness, then robust systems should exhibit a structure that reflects this trade-off. In fact, theoretical investigations of ecological systems (i.e., food webs) support this verbal model. Systems with completely modular structure are unstable (Pimm 1979a; Solow, Costello, and Beet 1999). However, several models have found increased robustness in ecological systems with intermediate levels of modularity using a number of different measures (Pimm 1979a; Rozdilsky, Stone, and Solow 2004; Teng and McCann 2004).

This chapter focuses on the role of control of flow and modularity in robustness. We do this in the natural context of networks, which are used to describe the interconnections among components within a system. Examples of the use of networks in ecology include how energy flows through the connections among species (e.g., from prey to predator in a food web; Paine 1980) and how organisms move among connected patches in the landscape (Keitt, Urban, and Milne 1997). In the context of social systems, examples of networks include patterns of information flow through social connections when a person searches for new job opportunities (Granovetter 1973) and distribution of power in exchange networks (Cook, Emerson, and Gillmore 1983).

We begin this chapter with a general introduction to networks and their use in social-ecological systems. A more specific discussion of the potentially important structural properties of networks for robustness in these systems forms the bulk of this chapter. The parts of this chapter focusing on social network analysis should be seen as complimentary to chapter 4. Whereas this chapter focuses on the structural properties of social networks, chapter 4 discusses social networks as a crucial component in successful natural resource management based on experiences from case studies.

WHAT CONSTITUTES A NETWORK?

NODES AND EDGES

In practice, a network is often represented as a graph, and there is an entire branch of mathematics, graph theory, that deals with the analysis of these graphs. Graph theory is particularly useful to us because it gives us a way to measure the structural properties of a network that we hypothesize are important for robustness (i.e., control of flow and modularity). Graphs consist of nodes and edges. Nodes are the terminal points or intersection points of the graph (sometimes called vertices). Edges are the links between nodes and represent the structure of the network over which interaction occurs. Interaction, or flow, along edges can be in either or both directions (see figure 3.1).

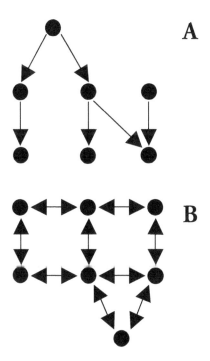

FIGURE 3.1. A) A unidirectional graph that could represent a hypothetical food web network. Nodes represent species and edges represent the interactions between species. The arrows indicate the direction of the interaction. B) A bidirectional graph that could represent a social network. Nodes represent farmers in a village and edges represent business relations.

More concretely, food webs are a type of ecological network where nodes correspond to different species and the edges describe the structure over which energy and/or nutrients flow in the ecosystem. Each edge or link in the food web represents who eats who. For a hypothetical example, see figure 3.1a. Networks are also used in ecology to represent landscape connectivity. Here, nodes are suitable patches of habitat and edges represent connections in the landscape between patches over which organisms flow (e.g., disperse).

Within social systems nodes usually represent individuals or organizations, but have also been used to represent resources and tasks, as used by Carley and Krackhardt (2001). Edges in a social network describe the structure over which information, influence, or resources generally flow. Each link is an abstraction of the social relations between the social entities represented by the nodes, such as relatedness, friendships, or business associations. Thus figure 3.1b could represent a hypothetical example of a social network where the nodes represent individuals (e.g., farmers) in a small village and the links represent bidirectional business relations.

WEIGHTS OF EDGES DESCRIBE DIFFERENCES AMONG INTERACTIONS

The edges of a network as a whole describe the topological structure of the system. The topology can have a significant effect on the system's behavior. However, the topology of the network alone may not completely capture how nodes' interactions shape the system. The topology of the network is the relationship described among nodes when all edges are assumed to be identical (e.g., dichotomous, an existing edge is set to 1, nonexisting edges are set to 0; see figure 3.1), but the interactions among nodes described by edges are unlikely to be identical in ecological and social systems. Topology, both alone and with consideration of weighted edges, provides information on different aspects of the network structure and its potential impact on the overall system's behavior. Ultimately, results regarding how network structure influences robustness may differ depending on whether or not edges are weighted (e.g., Krause et al. 2003).

Edge weightings reflect differences among interactions such as quality, frequency, or the level of intimacy (e.g., friends versus business associates

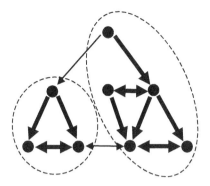

FIGURE 3.2. A hypothetical food web network. Nodes represent species and edges represent the interactions between species. The weights indicate the strength of interactions and the arrows indicate the direction of the interaction. Modules are encircled by dotted lines.

in social networks). More generally, these properties are often thought of as corresponding to the strength of interactions. Visually, the strength of interactions is usually expressed either as the length of the line connecting two nodes or by its thickness (see figure 3.2). It is possible to express multiple properties of interactions that contribute to the general strength of interactions by employing length, thickness, color, etc. For example, in this way, it is possible to include information both on the quality and frequency of interactions.

In food webs, measures of interaction strength could be the change in abundance of species A when species B is removed (functional edges; Paine 1980) or the amount of energy flow between species A and species B (energetic edges; Milne and Dunnet 1972; Baird and Milne 1981). Interaction strength in the landscape ecology context usually refers to the distance between patches of habitat (Urban and Keitt 2001), but has also incorporated the quality of the surrounding landscape matrix (Bunn, Urban, and Keitt 2000). Interaction strength in social networks could represent the frequency of interactions (e.g., how often does person A discuss important matters with person B), the type of interaction (e.g., kinship versus workplace contacts), or the level of intimacy. However, most analytical methods in social network analysis are based on the assumption of single-valued edges, even though some of the measures have been extended to incorporate weighted edges as well (see e.g. Wasserman and Faust 1994 for a review).

THE STRUCTURAL APPROACH

NETWORK ARCHETYPES IN SOCIAL AND ECOLOGICAL SYSTEMS

There are several archetypical network topologies (i.e., all interactions are weighted equally) that differently influence system behavior and robustness. Most of these networks have been found in social, biological, or engineered systems, but appear to be quite modified in ecological systems such as food webs. Perhaps the most basic network structure is the random graph. In a random graph each pair of nodes is connected with probability, p, which determines the exact structure of the graph (see figure 3.3a). It is well-known that random graphs do not capture well the structure of many real-world networks. However, they do exhibit the small world effect of many social networks, but without accurately capturing the distribution of connections of most real-world networks.

SMALL WORLD NETWORKS   In the 1960s Milgram sent letters person to person that reached a target individual, unknown by the original letter receivers, in around six steps (Milgram 1967). This result is termed the small world effect and suggests that individuals unknown to each other in social systems are connected on average by as few as six steps (e.g., six degrees of separation). In general, a specific network structure is termed a small world network if a relatively short path through the network connects any pair of nodes. Generally, the small world network (Watts and Strogatz 1998) begins as nodes on a one-dimensional lattice. Initially, each node is connected to each of its neighbors, as in figure 3.3b. The small world network is formed by rewiring the relationships between neighbors as one end of each edge is moved to a new location with probability p as in figure 3.3c. There are several other modifications of the small world network (Monasson 1999; Newman and Watts 1999), but in general the small world network sits somewhere between a random graph and a regular lattice. Its main property is that the number of steps between any two individuals is low. This property appears to be common in many real networks; however the distribution of connections in the Watts and Strogatz small world network does not match real networks well. The small world effect appears to be relatively common among social systems (Barabási 2002). Data from ecological systems, at least for food webs, suggest that formal small world networks are rare, but that the small world effect is common. That is, paths between nodes are even

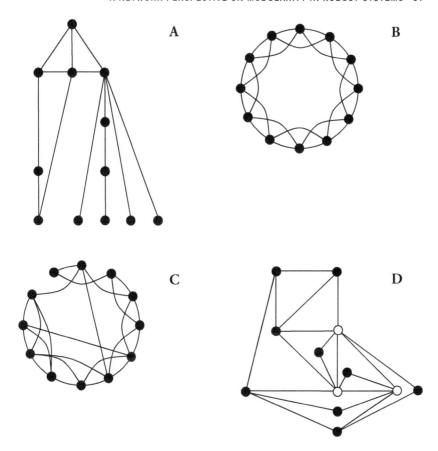

FIGURE 3.3. A) A random graph with twelve nodes. The probability of an edge between any two nodes is 0.2. B) A regular ring graph with twelve nodes. Each node is connected to its four closest neighbors, giving each node degree 4 (after Watts and Strogatz 1998). C) A small world graph constructed from the regular ring graph in B. Edges in the regular ring graph were rewired to a random node with probability 0.2. Duplicate edges were not allowed (after Watts and Strogatz 1998). D) An approximately scale-free network of twelve nodes constructed following a growth model where new nodes attached to two previously existing nodes. The three nodes with the highest degree (all with degree 6) are in white to highlight the existence of hubs in a scale-free network. The small size of this network prevents it from following a power-law degree distribution very closely

shorter than in social systems; on average approximately two links appear to connect pairs of nodes (Williams et al. 2002; Dunne, Williams, and Martinez 2002a). Landscape networks may also exhibit some small world properties (Brooks 2006).

SCALE-FREE NETWORKS   The degree of a node is defined as the number of edges connected to it. Like the small world effect, the distribution of node degree (degree distribution) is also structurally informative. In random graphs, for example, the degree distribution is binomial or Poisson for very large graphs (Newman 2003). A number of recent and ongoing network studies are focused on node degree distributions of large and complex real-world networked systems such as power grids, the World Wide Web, and scientific authors referring to each other's articles (Newman 2003). This research suggests that the degree distribution in most real networks is not random.

Many networks appear very complex, yet have an underlying order and follow simple laws (Barabási 2002). An important discovery within this field is the abundance of networks following power-law degree distributions (often called scale-free networks; see figure 3.3d), where most of the nodes have few links to other nodes, and the number of nodes declines exponentially with increasing node degree (Barabási and Albert 1999). This means that the degree distribution is highly skewed with few nodes having large numbers of connections. These networks suggest many characteristics of interest to robustness, such as the importance of the few nodes with many links (e.g., hubs), since these well-connected nodes tie the network together, preventing it from being fragmented in many disconnected subnets. Also, scale-free networks exhibit the small world effect described above. Many networks, including social and biological systems, have been shown to exhibit scale-free degree distributions. Numerous models explaining how scale-free networks form have been proposed; as an example, see the list in Barabási, Ravasz, and Vicsek (2001). Common features of processes in models that result in scale-free networks are incremental growth and preferential attachment. Incremental growth occurs when networks are assembled through the addition of new nodes to the system (see figure 3.3d), while preferential attachment encodes the hypothesis that new nodes connect with higher probability to more connected nodes (Barabási, Ravasz, and Vicsek 2001), e.g., the rich get richer (*sensu* Simon 1962).

Ecological networks differ from social networks and do not generally exhibit scale-free degree distributions, although their degree distributions are still skewed toward few nodes with many connections. Few nodes with many connections (e.g., hubs) have been observed in food webs (Dunne, Williams, and Martinez 2002a), landscape networks (Brooks 2006), and epidemiological networks (Woolhouse et al. 1997; Meyers et al. 2003). Landscape networks often arise from habitat destruction creating a network of patches from formerly contiguous habitat (Melian and Bascompte 2004; Fortuna and Bascompte 2006). Thus, landscape networks, at least, are neither as-sembled by the addition of new patches nor grow via dispersal from the most connected patches. Despite the existence of hubs, landscape networks are unlikely to be truly scale-free because of the processes by which they are built. A set of biological networks that do appear to exhibit scale-free dis-tributions is metabolic pathways and regulatory networks. Food webs and landscape networks differ from these other biological networks in at least one significant way: selection can act to optimize metabolic and regulatory networks at the whole system level, but this is not the case for food webs and landscapes. In these systems structure is an emergent property of the system where optimization is at the level of its components or lower, and perhaps this is the reason why topology in networks representing ecosystem processes appears to differ from many other types of networks.

BENEFITS OF STRUCTURAL NETWORK APPROACHES

Central in theories of robustness in CAS are emergent, global patterns and properties arising from local interactions (Levin 1998). Networks can be viewed as maps that outline these local interactions and preserve their importance at the system level. Because of this preservation, network ap-proaches are often an improvement over approaches that average over all the interactions that an individual actor or species participates in (e.g., mean-field models). By incorporating local interactions, network approaches are able to represent the structure of the interactions, and the emergent proper-ties contributing to robustness in CAS partly depend on this structure.

These arguments go hand in hand with the assumption in the social sciences underlying the study of individual and collective behavior: the im-portance of the context in which the actions take place (Gladwell 2002). A fairly well-established assumption among social scientists is that individual behavior and opinions are rooted in the structure to which people belong (Degenne and Forsé 1999). The structure itself can be described as a net-work of relations among individuals, i.e., social networks. The concept of the social network has been used in a number of disciplines to explain phe-nomena and build theories, and the structural approach was initiated as early as a century ago (Freeman 2004). It can enrich economic theories in cases where the historical assumption of freely and easily available infor-mation is not upheld, but instead, information as well as social influence is channeled through social networks (Abrahamson and Rosenkopf 1997). The role of the entrepreneur's social network is emphasized in theories of the emergence of new organizations and communities of organizations (Aldrich 1999). Far-reaching networks of organizations such as transnational corporations are also of importance in explaining social change and globalization (Ahrne and Apostolis 2002).

Networks have long been used in ecology to represent food web data (Elton 1927). The links between species are fairly readily observable in nature, and their representation in a network format provides a description of community structure (Paine 1980). This structure is thought to be linked to community stability (MacArthur 1955), although more formal analyses of the relationship between network structure and robustness are relatively recent (Dunne, Williams, and Martinez 2002b; Garlaschelli, Calarelli, and Pietronero 2003; Krause et al. 2003). The use of networks to represent dispersal among habitat patches is also relatively recent (Urban and Keitt 2001; Brooks 2006). In this context information on landscape structure derived from the network is used to identify individual patches that are important to landscape level connectivity. This usage of networks and graph theory has critical application to conservation biology.

DEFINITIONS OF ROBUSTNESS IN A STRUCTURAL CONTEXT

Definitions of network robustness may differ from more general notions of resilience described elsewhere in this book. From a purely structural perspective, e.g., dichotomous nodes and edges, a common definition of robustness is the vulnerability of the network to defragmentation. Continual removal of nodes and/or edges will eventually fragment a network into components (i.e., isolated subgraphs) or, failing that, make distances between nodes so large as to be practically disconnected. Hence, one commonly used definition of a network's robustness is its ability to withstand removal of nodes or edges without fragmenting into disconnected components. However, this concept of robustness comes with an ambiguity; namely that robustness to node and/or edge removal depends on how nodes or edges are removed. The overall distinction is between random removal and targeted removals following some strategy. For example, scale-free networks are robust to random removal, but vulnerable to targeted attacks where highly connected nodes are removed first (Albert, Jeong, and Barabási 2000).

Various network structures have been suggested as more or less tolerant to targeted attacks; for a theoretical example see Shargel et al. 2003. It also appears that the defragmentation process is often highly nonlinear, i.e., the breakdown of the networks occurs rapidly (see e.g., Urban and Keitt 2001; Dunne, Williams, and Martinez 2002a; Bodin et al. 2006).

A plausible extension to defragmentation is to include congestion (e.g., assuming an edge carries flow of some resources and individual nodes receiving too many resources per time unit can experience congestion). Thus, the robustness of a network can be defined as 1. its capacity to withstand removal of nodes without being fragmented into smaller pieces and/or 2. its capacity to uphold the flow of resources, e.g., to limit individual nodes experiencing congestion (Holme 2002; Dodds, Watts, and Sabel 2003).

DETERMINING MODULES AND GROUP MEMBERSHIP

Within a network there may be groups of nodes (i.e., modules) that according to some criteria distinguish themselves from the rest of the nodes in the network (see figure 3.2). One such criterion is called cliques in social science, where the distinction of groups is based on cohesion, i.e., a high density of interconnecting links among each clique's members. The original definition of a clique, assuming nondirectional relationships (e.g., pairs of reciprocal edges), is a subset of nodes that are all adjacent to each other and where there are no other nodes that are adjacent to all of the members of the clique (Wasserman and Faust 1994 and references therein). This definition can be extended to account for directional and continually valued edges as well. The definition of a clique is rather strict, e.g., every node must be adjacent to every other node in the clique. To include intermediate paths, e.g., allowing "stepping-stones" to provide a way to connect clique members, the definition of n-cliques can be applied (Alba 1973). Also, group membership can be asserted based on a minimum number of links to other members (e.g., k-plexes; see Seidman and Foster 1978).

Both n-cliques and k-plexes are important characteristic of cliques. A particular node can belong to several cliques at once. If one wants to view these groups as separated units or compartments, multiple memberships of individual nodes limits the applicability of the idea of separate units. LS-sets and Lambda sets solve this possible problem, since they are based on intra-group link densities and generate separate branches of hierarchical clusters of groups (see Borgatti, Everett, and Shirey 1990 and references therein). However, multiple memberships can also be seen as an important characteristic of networks, where nodes belonging to several groups could be interpreted as bridges. In social systems such analysis can reveal hints about how individuals influence others through multiple group membership (Everett 1998). Also, there appears to be underlying commonalities between various kinds of networks where the concept of overlapping groups applies (Palla et al. 2005). In general, the existence of overlapping groups sets up the potential for intermediate levels of modularity within the network.

Within ecological systems cliques are often referred to as modules or compartments and are based on the density of links (Dunne, Williams, and Martinez 2002a) and sometimes the weight of links as well (Krause et al. 2003). In food webs modules are formed by sets of strongly interacting species that overlap other modules through weak connections among species (McCann, Hastings, and Huxel 1998). In the landscape context, modules are formed by habitats that are relatively easy to migrate among (Urban and Keitt 2001). These modules become coupled by long distance or rarer migration events. A network structure with slightly overlapping cliques or compartments is one in which the mechanism of intermediate modularity might act to enhance robustness.

Another way of distinguishing groups within a network is to cluster individuals according to the set of relations they have with one another. This type of clustering is called equivalence and is based on the clustering of individuals possessing the same type of structural relations to others (Degenne and Forsé 1999). Equivalence analysis can help to find different roles or classes (like managers-workers or teachers-students in social systems and functional group or trophic level classification like predator-prey in ecological systems) and could also give insights into different power structures and the control of flow of information or resources. The definitions of equivalence (e.g., regular and structural) are quite rigid; real-world complexity and diversity of links sometimes makes it necessary to settle for statistical approximations of equivalence (Degenne and Forsé 1999).

Technically, hierarchical clustering procedures are commonly used to distinguish clusters (e.g., groups) that are unique yet internally homogeneous. The similarity of nodes within the cluster can be based on cohesion, equivalence, or other similarity measures. These techniques are quite general. For example, methods developed in the social sciences have been applied to food webs to determine the existence of modules (Krause et al. 2003).

THE IMPORTANCE OF NODE POSITION WITHIN THE NETWORK

Having seen how network analysis can be used to find different groups, we now turn to the network positions occupied by individuals to understand their possible structural effect and role in controlling flow within the network.

An individual's social capital is influenced by the network structure the individual is embedded in. Social capital, as expressed in this individualistic structural approach, consists of an individual's personal network and her chances of satisfying her needs while participating in her personal network. However, the concept of social capital is not solely the volume of contacts, but depends also on the structural characteristics of the relations (e.g., not all relations have the same value). In addition, not only are direct connections accounted for, but indirect connections are of great importance (Granovetter 1973). Related to the social capital (or the power or influence) of an actor is the redundancy of its contacts. Here, redundancy is used in the sense that there are alternative paths for others to reach a particular actor in an individual's personal network. A nonredundant link is assumed to be more valuable than a redundant one for an actor possessing it, since the actor will have exclusive access to the actors connected with a nonredundant link (Burt 1992). Such links ("structural holes") can be particularly valuable when they connect different groups. Different archetypical roles can be asserted to such actors such as gatekeepers, brokers, and coordinators (Gould and Fernandez 1989). However, as seen from a network structural perspective, structural holes decrease the robustness of the network. If nodes possessing nonredundant relations are removed, the network experiences defragmention.

The story within food webs is a parallel one for the most part. Instead of social capital, species must satisfy their energy demands through their local interactions in the network. Again, energy demands are satisfied not only by the number of contacts with resources or prey, but by the strength of the interactions (e.g., the level of specialization on a particular prey species). Overall, both direct links and indirect links within the food web are important because together they determine the importance of a species in maintaining the structure of the food web. The structure of the network is likely to depend fundamentally on highly interactive species (Paine 1980; Raffaelli and Hall 1996; Soule et al. 2003). The concept of a highly interactive species is related to the idea of centrality discussed below. Within food webs, redundant pathways that link consumers to basal resources retain value because redundant links allow consumers to maintain their connection to fundamental energy sources even if some pathways deteriorate due to the depopulation or extinction of species intermediate in the pathway (Dunne, Williams, and Martinez 2002b; Garlaschelli, Calarelli, and Pietronero 2003). Similarly, within landscape networks, nodes (patches) with many connections (e.g., hubs) are important in determining the pattern of dispersal and resource use by organisms (Urban and Keitt 2001; Brooks 2006). The importance of such nodes echoes our previous discussion of hubs, targeted removal of nodes, and robustness of the network.

A fundamental trait of a node's position is centrality. Numerous studies in social science agree that power is bound up with centrality, although the connection is not completely straightforward (Degenne and Forsé 1999). This is echoed in the ecological literature through the importance of highly interactive species and habitat patches with many connections. Several perspectives of centrality are possible; here are some of them:

1. Degree centrality (Freeman 1979) is the number of connections a node possesses. i.e., a node with lots of connections, compared with the other nodes, has a relatively high degree of centrality. Also, the whole network can be characterized by a centrality index.

2. Closeness centrality (Sabidussi 1966) aims to weigh a node's connections to other nodes within the network, i.e., how close it is to the others. As above, the whole network can also be characterized by a closeness index.

3. Betweenness centrality (Freeman 1979) aims to weigh the number of pathways that run through a specific node; it estimates a degree of in-betweenness of a node. If connections are weighted, there are similar measures that focus on how "costly" it is to run a pathway through a specific node (e.g., flow betweenness in social systems). Once again, the whole network can also be characterized by a betweenness index.

4. The concept of keystone (Paine 1969; Power et al. 1996) from ecology is defined as a species (i.e., node) that has a stronger effect on the overall network, usually due to a high number of connections, than would be expected based on its abundance, i.e., a rare species with many connections. Dunne, Williams, and Martinez (2002a) have also suggested the idea of a structural keystone, i.e., a species that has a stronger effect on the overall network, usually due to its position, than would be expected based on the number of connections it has.

The link between power, prestige, and influence on one hand and centrality on the other is, as already stated, not completely straightforward. By using several measures and perspectives, a more complete view of a node's or a group's influence can be estimated. Network structure analysis has also been used both for intraorganizational and interorganizational dependencies assessment. Similarly, in the food web and ecosystem context the relationships among centrality, keystone, and highly interactive species are not straightforward (Paine 1969, 1980; Raffaelli and Hall 1996; Soule et al. 2003).

THE QUALITY OF EMPIRICAL NETWORK DATA

COLLECTION AND QUALITY OF ECOLOGICAL NETWORK DATA   In many senses collection of food web data is relatively straightforward. The species in a community are usually readily observable and the interactions among species may be relatively easily observed as well. However, sometimes the methods of observation required are very tedious to perform, e.g., gut analysis or stable isotope analysis. From these observations a network topology can be constructed that reasonably describes the community structure (Paine 1980). However, that is not to say that food web data are without problems. Data for constructing food webs are somewhat limited for taxa, interactions, and even complete functional groups (such as bacterial decomposers) that are not easily observed. An alternative is to focus on a few species in the community and actively search out all of their interactions. This focused approach gives more complete and detailed information, but is often only done for a portion of the entire system. By either approach, there is often unequal resolution of the food web that can create bias in analyses (Dunne, Williams, and Martinez 2002a). One way of dealing with this is to aggregate species in the food web into trophic taxa by grouping together species that share the same set of predators and prey into a trophic species (Dunne, Williams, and Martinez 2002b).

While determining the topology of the network may be relatively straightforward, determining the edge weights is not. Functional food webs have edge weights that measure the per capita influence of an individual species on the community. This requires experimental manipulation of the abundances of the species whose influence is to be measured (Paine 1992). Energetic food webs have edge weights that measure the amount of energy or material flowing from lower trophic levels to higher trophic levels. These are usually determined from literature values, physiological experiments, or the use of stable isotope tracers. In addition to the difficulty of obtaining these measures, there are a number of other problems. First, these measures of interaction strength often do not have the same meaning as parameters used in models of species interactions. This makes it difficult to move between the predictions of models (e.g., Lotka-Volterra predator-prey model) and food webs. This is a problem because it is often easier to investigate and quantify stability and response to disturbance in models than in food webs—important information for robustness. Additionally, these different indexes affect our interpretation of the ecology of a system because functional and energetic food webs of the same system may have different topologies. Edge weights that are not per capita measures make it difficult to compare what is happening at different localities and over time (Berlow et al. 1999). This is particularly a problem in robustness where the strength of interactions is important because edges that are weak in a broad spatial or temporal context may be quite strong locally or over shorter, yet key, periods of time (Berlow 1999).

For landscape networks, geographical information systems (GIS) data are often used (Urban and Keitt 2001). These data provide such information as elevation, temperature, vegetation in association with spatial coordinates that can be used to reconstruct type and location of various habitats. These data are increasingly gathered using remote sensing techniques, but much of the GIS information still comes from map data.

COLLECTION AND QUALITY OF SOCIAL NETWORK DATA   Research on social networks requires consideration of several issues that influence any attempt to empirically assess network structures. Here some of those issues are briefly discussed.

Defining the boundaries of the population (the nodes) within which the interactions of the population (the edges) are to be measured is in many cases nontrivial. In some cases the population is well defined a priori, for example, if one is measuring the relations between farmers in a certain village. Here the population is determined in a geographical context, i.e., an attribute (the address) of the individual determines whether or not he or she should be included or not. In other cases, for example, when measuring a network of friends originating from a few initial "seed" individuals, the population of "friends" is by nature undetermined until relations are assessed.

The ability to determine the population prior to the relational data gathering process often imposes constraints on the data gathering process itself.

Further, one has to distinguish between studies where all relations in the population are to be enumerated, e.g., assessing the complete network from studies where samples of individuals are to be drawn from the population. The latter, where samples of individuals' self-centered networks, e.g., egocentric networks, are measured aims for assessing general network characteristics such as average network density of the complete network. The appropriate choice of sampling technique depends on a number of issues such as population size and expected number of relations (see Frank 1981 for an example).

Fundamental in social network analysis is the nature of the social relations (i.e., the edges). Asserting the content (and existence) of relations is a research area in itself. An empirical study should include rigorous methods on what kind of relations one is looking for as well as how to methodologically proceed to evaluate and find them. Several studies have revealed that respondents exhibit limitations in remembering interactions with others, and there could be a rather big difference between self-reported relations and actually existing relations as assessed by observations (for a survey, see Marsden 1990). This holds true especially for less frequent or intense relations. However, those limitations can normally be addressed, given the researcher is aware of them. Also, sometimes one wants to distinguish between actual relations and perceived ones (as discussed by Krackhardt 1987). For example, in diffusion processes actual relations are of interest; when studying influence, perceived relations are of interest. Finally, measuring relational data by observation is very time and resource consuming; hence most social network studies rely on self-reports using questionnaires and/or interviews.

As stated above, defining population and/or network boundaries can impede and constrain the data gathering process. Assuming that the population is known a priori, the researcher can compile a list of respondents and then, assuming a whole network approach, ask each and every one on the list about his or her relations with the others on the list. This can be done either using recall methods (i.e., the respondent generates a list of his or her relations) or using recognition methods (i.e., the respondent picks individuals from the precompiled list whom he or she relates to). The latter normally captures more relations (Brewer and Webster 1999).

If the population is not known a priori, a precompiled list of the population is not accessible. Hence the researcher has to resort to methods to determine, or define, the population (and hence the network) boundary. The researcher can use informed experts to compile such a list. Another method is to resort to the snowball survey, where one starts with a sample of nodes and then maps their personal star network (i.e., egocentric network). Then one goes on repeating the process with the newly obtained nodes until an appropriate number of nodes are collected (Goodman 1961). The snowball survey has been shown to provide the researcher with a more comprehensive estimate of the actual network population than using informed experts (Doreian and Woodard 1992).

RESEARCH ON ROBUSTNESS IN SOCIAL AND ECOLOGICAL SYSTEMS USING NETWORK ANALYSIS

MODULARITY IN ECOLOGICAL SYSTEMS

The prediction that intermediate levels of modularity should enhance robustness was proposed in the context of food webs in the early 1970s by May (1972, 1973). Later theoretical investigations have generally supported this notion (Pimm 1979a, b; Rozdilsky, Stone, and Solow 2004; Teng and McCann 2004). In the last thirty years data has slowly amassed to the point where it is possible to begin looking at the prevalence of modular structure in real food webs. The general expectation is that the signature of modularity should be visible in food webs if it enhances robustness because extant systems have generally been robust to disturbance in the past. The structure of intermediate modularity is created by modules of species with strong functional interactions that are then weakly coupled to one another. Overall, empirical food webs do have a preponderance of weak links (Paine 1992; Goldwasser and Roughgarden 1993; Fagan and Hurd 1994; Raffaelli and Hall 1996; Wooton 1997; Berlow et al. 1999) that are a hallmark of intermediate levels of modularity. Direct analysis of the structure of empirical food webs is suggestive of intermediate modularity, but different methods of analysis have provided inconsistent evidence for its being the dominant pattern (Dunne, Williams, and Martinez 2002b; Krause et al. 2003; Allesina et al. 2005). Dunne, Williams, and Martinez (2002b) analyzed sixteen empirically derived food web data sets from a topological perspective. Of these, they found five food webs that exhibited clustering, and these were the webs with the lowest connectance (connectance is defined as the actual number of connections among nodes divided by the total number of possible connections.). Interestingly, these same food webs exhibited small world topologies and scale-free degree distributions, which are unusual network properties for food webs. Krause et al. (2003) found that the ability to detect modularity depends also on the level of resolution in the food web and whether or not edge weights are included in the analysis. Only one of fourteen food webs originally analyzed showed modularity, but when the level of resolution was enhanced, three out of five food webs exhibited modularity. Krause et al. also found two food webs where modules were discernible only when the strength of interactions (edge weights) was included in the analyses.

In addition to the CAS robustness mechanism of intermediate modularity discussed throughout this chapter, the weak links inherent in a topology with intermediate modularity may stabilize systems. It is not clear whether or not this stability ultimately translates into robustness, but internal stability may impact the ability of a system to respond to outside disturbance as well. Ecological models based on empirical food web data that reflect the ubiquitousness of weak links suggest that these weak links can stabilize oscillatory dynamics (McCann, Hastings, and Huxel 1998). Essentially, weak links can couple pairs of strongly interacting species that alone exhibit oscillatory dynamics with other species or groups or species that do not oscillate. The weak coupling of the oscillatory group to the stable group acts to reduce the amplitude of oscillations in the oscillatory group and overall produces a more stable system.

The evidence for intermediate modularity as a mechanism to reduce spread of disturbance and directly impact robustness is mainly from anecdotal evidence and models. Direct evaluation of the importance of this mechanism for food web robustness will require long-term food web data for multiple systems with and without modules that have experienced disturbance at some point during their monitoring. Krause et al. (2003) simulate the effects of disturbance on the Chesapeake Bay food web. This food web consists of two modules corresponding to pelagic and benthic species. Overharvesting is simulated by removal of a pelagic fish, which results in the loss of other species. The decrease in the density of links was higher in the pelagic module than in the overall system, and there was no change in the benthic module, suggesting that the effects of overharvesting were limited to only the module that had been disturbed. This is consistent with the predictions of the hypothesis that intermediate modularity should stop the effects of disturbance from propagating throughout the entire system.

Another picture of the importance of intermediate modularity is found in a spatial context. Here, the primary interest is modular spatial structure formed by individuals that aggregate spatially. Intuitively, spatial modularity should be important for the robustness of ecological systems impacted by disturbances that spread spatially. Related hypotheses have also been raised in the context of risk spreading (den Boer 1968; Urban and Keitt 2001) and small world networks (Watts 1998; Watts and Strogatz 1998). Simulation models suggest that intermediate spatial modularity may enhance the robustness of populations (Webb 2008). When robustness is measured as the minimum population size following a spatially spreading disturbance, a distinctive optimum appears at intermediate levels of aggregation in the spatial distribution of organisms. This is the result of a trade-off between locally high levels of spatial connectance with strong aggregation and globally high levels of spatial connectance with little aggregation. When dispersal is the mechanism that generates the spatial distribution, populations with intermediate spatial modularity may not pay much of a penalty in terms of their realized net reproduction. Small differences are expected in net reproduction with different probabilities of long-distance dispersal, because propagules that disperse long distance are more likely to find open space than those that disperse locally. Because of the similarity in reproductive rate, we see little trade-off between engineering resilience (the return time of the system) and the minimum population size postdisturbance.

CONTROL OF FLOW IN ECOLOGICAL SYSTEMS

The role of control of flow in robustness can be examined by considering the importance of the position of nodes and edges in the network and their relationship to how disturbance spreads in the system. The idea that the position of a species in the ecosystem may be important is driven by several examples of this in the food web literature. Perhaps the most famous is the cascading effects in Pacific coastal systems following the decline of the sea otter from intensive hunting for the fur trade (Estes and Palmisano 1974; Steneck et al. 2002). Sea otters are important predators that control sea urchins, which are important grazers on algae. With sea otters in place,

Pacific coastal systems in North America look like underwater kelp forests. When sea otters were removed from the system, urchin populations exploded, destroying kelp forests and producing a cascade of effects for species dependent upon the kelp. With protection for sea otters, their populations have again increased, and kelp forests and their associated species are also bouncing back where sea otter populations have reestablished (Estes and Duggins 1995). The importance of the position of the sea otter in regulating Pacific coastal ecosystems is recognized, with sea otters thought to be a keystone species (Paine 1969).

General network theory suggests that certain positions and edges in a network are important in small world and scale-free networks. However, food webs only fit these archetypes to a limited degree. Are the same positions and edges important in food webs as in the archetypes? Like scale-free networks, food webs are more robust to the random removal of nodes than the selective removal of species with the most links (Solé and Montoya 2001; Dunne, Williams, and Martinez 2002a; Allesina and Bodini 2004). Here robustness refers to the number of secondary extinctions caused by the removal of a node. Extinction occurs when a nonbasal species loses all of its prey species. Overall, the behavior of food webs is much like that of scale-free networks, likely because food webs do have skewed degree distributions. Those food webs with scale-free degree distributions were the most vulnerable to attack, but, as is discussed below, this is most likely because these food webs also had the lowest connectance.

The number of secondary extinctions as a measure of robustness is related to defragmentation of the network. The work of Dunne, Williams, and Martinez (2002a) also found that there were some species with few links whose removal caused a surprisingly large number of secondary extinctions, i.e., structural keystone species. Structural keystone species likely play an important role in maintaining the minimum spanning tree in the network. A minimum spanning tree is the shortest path through the network that connects all of the nodes together. Consistent with the idea that maintenance of the minimum spanning tree is essential to robustness, Dunne, Williams, and Martinez found that robustness increased with the level of connectance. In fact, there appears to be a threshold effect where highly connected webs are relatively insensitive to removal of species up to when about 20 percent of species are removed. Beyond this the number of secondary extinctions greatly increases, suggesting defragmentation has begun because the minimum spanning tree has been cut. Highly connected webs have a greater redundancy of links that protects the minimum span-ning tree from defragmentation during node removal (Garleschelli et al. 2003; Melian and Bascompte 2004). In this context particular edges be-come important as redundant links enhance stability and links in the mini-mum spanning tree enhance the efficiency of energy transfer through the food web (Garleschelli, Calarelli, and Pietronero 2003). Certain equiva-lence classes, particularly basal species, are also important in maintaining the minimum spanning tree and can be thought of as structural keystone species (Dunne, Williams, and Martinez 2002a). Interestingly, although the whole food web does not follow a power law degree distribution, the minimum spanning tree itself does appear to (Garleschelli, Calarelli, and Pietronero 2003).

In landscape ecology three different types of habitat patches (nodes) are recognized: those that contribute to recruitment of a species, flux in species abundance, and traversability of the landscape (Urban and Keitt 2001). The relative importance of these equivalence classes will depend on the specific system and disturbances, but the network perspective suggests that conser-vation efforts should be concentrated on the minimum spanning tree be-cause it allows dispersal over the entire landscape (Urban and Keitt 2001). Keitt and colleagues (Keitt, Franklin, and Urban 1995; Keitt, Urban, and Milne 1997) suggest that two types of patches are important in maintaining the minimum spanning tree: core patches and stepping-stone patches. Ur-ban and Keitt (2001) discovered a similar threshold to that found in the food web studies when studying robustness of the spatial network by removal of edges corresponding to links between habitat patches. The network moves very quickly from connected to fragmented over a narrow range. The length of the pathway through the spatial network initially increases as di-rect connections between distant nodes are lost. These direct connections are replaced by stepping-stonelike paths, but as these stepping-stones are lost from the network, the network fragments. Core patches maintain the minimum spanning tree, but stepping stone patches provide redundancy in the connections between distant parts of the network. An example using the dispersal patterns and habitat of the Mexican spotted owl suggests that when habitat patches are removed under a number of schemes, occupancy of habitat by the owl declines, but if the minimum spanning tree is pro-tected, then occupancy can be maintained despite removal of patches (Ur-ban and Keitt 2001). It is hypothesized that this is because the traversability of the landscape is maintained, e.g., the owl can disperse throughout all the habitat patches potentially available to it. A similar example comes from southern Madagascar where a network of forest patches would rapidly defragment following destruction of the smallest, sometimes falsely perceived as insignificant, patches (Bodin et al. 2006).

NATURAL RESOURCE MANAGEMENT

The theme of this book is sustainable natural resource management in coupled social and ecological systems. Following the examples above of the network structural approach in ecological systems, the next step is to examine the network structural approach in social systems. Here key topics from current research in integrated natural resource management are briefly revisited with attention directed to social networks. The objective is to single out the importance of social networks as one possible common ground for those theories. In the following chapter by Hahn et al., arguments based on empirical studies are presented on the critical importance of social networks in successful natural resource management. Hence this section, a review of key topics in natural resource management, should be seen as complimentary to the following chapter.

Adaptive management focuses on continual change in management practices following increased knowledge of the system's behavior. Much emphasis is put on stakeholders at different levels and their involvement in the management process. Knowledge and information from the natural system's responses to (current) management practices are critical assets that must be transferred through communication links on different levels to provide for evaluation and possible adjustments of practices in use. Hence the function and structure of those communication links (i.e., networks) are both structuring and supporting the policy system. General structural characteristics of the network that could enhance understanding and assessment of functions and robustness are of interest, for example, the degree of centrality gives hints to the degree of diversity and decentralization of information as well as to the control of flow. Assessment of the network's geometrical distance could be used to determine the accessibility of information as well.

In adaptive comanagement (Gadgil et al. 2000; Olsson 2003), which to some extent can be seen as an extension of adaptive management but with a greater focus on local or community-level initiatives, emphasis is put on the bottom-up approach of natural resource management. Different actors and groups across all levels of organization are intended to comanage the environment, and accordingly the relations within and across organizational and/or group boundaries are of crucial importance (see, e.g., Carlsson and Berkes 2005). In addition, making use of local ecological knowledge requires a smooth flow of information and knowledge among the actors that should preferably scale up from individual and group levels to community and society levels. The structure and functioning of such networks are accordingly fundamental for understanding and successful implementation of those policy systems (Olsson 2003; Redman and Kinzig 2003). Bodin and Norberg (2005) used a modeling approach to gain insight into the impact informational network structures have, both on the aggregated system level as well as on the individual managers' level when adaptively managing a complex multistate ecosystem. The results indicate that sharing information in a network is highly beneficial over a broad range of network structures. But if the network gets too dense—that is, almost everyone is connected to everyone else—the intense information sharing reduces individual variation and can lead to large-scale (i.e., system-level) ecological collapses. In other words, if the level of modularity is reduced on behalf of dense global interconnectedness, the system experiences a reduction in robustness following the reduction in diversified management experiences.

At the heart of self-organization and creation of institutions lies communication between individuals. In the study of common-pool resources, formation of self-imposed regulations of resource extraction is central to avoiding the tragedy of the commons and resource depletion. Relations and trust building among resource beneficiaries are a necessity (Ostrom 1990), and it follows that the structure and dynamics of social relations should be part of the study of common-pool resources. Going further into polycentric institutions (Ostrom 1998), the establishment of boundary-breaking social links between different organizational and institutional levels is central. In this field structural network analysis can give valuable insights into the potential for such policy systems as well as providing clues to the robustness of organizational/institutional linkages. An empirical field study by Schneider and colleagues (Schneider et al. 2003) shows an excellent example of such a research approach. Here social network analysis was successfully employed in studying community-oriented natural resource management. Relations between various governmental agencies on different scales, also including nongovernmental groups and organizations, were measured, and the resulting networks were analyzed with respect to a number of issues such as the degree of boundary-spanning relations between fairly homogenous groups. The results indicate that appropriate network constellations can bring together fragmented agencies, user groups, etc., to establish common rules and cooperation. Using the terminology presented in this chapter, intermediate modularity provided for developing consensus on several issues among a variety of groups and organizations. In addition, it seems like public authorities can facilitate the creation of such cross-boundary networks.

Within the theory of adaptive cycles, i.e., "panarchy" (Gunderson and Holling 2002), emphasis is put on a system's renewal following a crisis. The abilities of a society to respond to a natural resource crisis depends to a large extent on previous, somehow collectively stored, experiences (Hilborn 1992; van der Leeuw 2000; Folke, Colding, and Berkes 2003; Redman and Kinzig 2003). Those stored experiences can be called social memory (McIntosh 2000), and a number of required functions for such collective memory to be successful in natural resource management have been hypothesized (Folke, Colding, and Berkes 2003). These functions are actually roles distributed among individuals in a (social) network. Individuals possessing one or several of these roles, linked together in a network, are suggested as being the cornerstones for successful natural resource management (Olsson and Folke 2001; Olsson 2003). Thus the social network forms the base from which individuals act, but the network of individuals also provides resources such as social memory that individuals and groups can exploit. Successful natural resource management would then depend upon the inclusion of individuals and groups with the appropriate roles in a network with a structure that provides for collective resources such as social memory. Here social network analysis can help to reveal and locate key individuals and roles, for example, by using different measures of centrality. Individuals with a high level of centrality could hypothetically have a large impact on the control of information and knowledge flow, e.g., they occupy roles such as brokers and/or facilitators.

Drawing from graph theory, a large body of results about networks suggests archetypical networks and characteristics of these networks that could potentially be important in the study of robustness. These general results suggest that characteristics like the connectedness and centrality of individual nodes, the connectedness of the network overall, minimum spanning trees, redundancy of connections, and modular structure are particularly of interest. The real question becomes how this plays out in the less ideal and noisier networks inherent in natural ecological and social systems. Although no comprehensive studies exist, a growing body of anecdotal evidence suggests that these characteristics do influence robustness in a variety of different types of ecological and social systems. Because of this, we suggest that these characteristics should be some of the first to be explored in the study of robustness in coupled social-ecological systems and in adaptive management.

A network perspective can also be used to address questions of scale. Network structure has the potential to reveal interesting properties, both on aggregate (e.g., compartment) and individual levels of analysis. It enables analysis on different levels of abstraction (individual, group, spatial, etc.) so that the most appropriate level can be used for a given problem. For example, identification of different equivalence classes, positions of power because of connectedness may be useful in disseminating information or gaining knowledge from key individuals or groups in an adaptive management perspective.

It is important to clarify that neither ecological nor social network structures are fixed. We have chosen in this chapter to focus on network approaches, but, particularly in the ecological literature, a parallel dynamic systems approach has also given much insight into questions of robustness (Pimm 1979b, 1980; Borrvall, Ebenman, and Jonsson 2000; and Ebenman, Law, and Borrvall 2004). Network approaches are still being developed for ecological and social systems, yet alone coupled social-ecological systems, but attention has been drawn to the importance of analyzing dynamic networks that would combine these two approaches (Breiger, Carley, and Pattison 2003 and reference therein). In particular, static networks fail to capture the importance of individuals who indirectly mediate the interactions of other individuals (Borer et al. 2002). One way of inferring network dynamics is to repeat the types of structural analysis discussed in this chapter at several points in time, but this can be quite difficult and requires larger amounts of data (Degenne and Forsé 1999). Computer simulations can also be employed (Holme 2003). Previous attempts within the ecological literature at such dynamic network models have been somewhat controversial because of the simplifying assumptions that are necessary (Brose, Williams, and Martinez 2003; Kondoh 2003a, b). At present, structural network analysis, because of its static nature, does not give much insight into studying network dynamics (i.e., how networks are formed and evolve), but it can reveal

how existing groups and individuals may become connected or disconnected. Such analysis points to network structures that could emerge from
the current one. Understanding possible future network structure is important in determining potential future ecological and management scenarios
that management might actively want to work toward or to avoid (chapter 7)
whether network dynamics are understood exactly or are inferred from
static network analysis.

Networks are a representation of both ecological and social systems.
They have the strong advantage of addressing not only the members of the
system but their interactions as well. Thus both potentially permanent aspects of the system, such as members, and more ephemeral aspects, such
as interactions, can be illustrated and their relative importance appropriately weighted. Because networks represent ecological and social systems
relatively well, it is not a stretch to suggest that they might also be useful
in representing coupled social-ecological systems (see Janssen et al. 2006),
particularly since social networks have specifically been identified as important in the context of social-ecological systems (see chapter 4).

REFERENCES

Abrahamson, E., and L. Rosenkopf. 1997. "Social Network Effects on the Extent of Innovation Diffusion: A Computer Simulation." *Organization Science* 8:289–309.

Ahrne, G., and P. Apostolis. 2002. *Organisationer, Samhälle och Lobalisering.* Lund: Studentlitteratur.

Alba, R. D. 1973. "A Graph-Theoretical Definition of a Sociometric Clique." *Journal of Mathematical Sociology* 3:113–126.

Albert, R., H. Jeong, and A. L. Barabási. 2000. "Error and Attack Tolerance of Complex Networks." *Nature* 406:378–382.

Aldrich, H. 1999. *Organizations Evolving.* London: Sage.

Allesina, S., and A. Bodini. 2004. "Who Dominates Whom in the Ecosystem? Energy Flow Bottlenecks and Cascading Extinctions." *Journal of Theoretical Biology* 230: 351–358.

Allesina, S., A. Bodini, and C. Bondavalli. 2005. "Ecological Subsystems via Graph Theory: The Role of Strongly Connected Components." Oikos 110:164–176.

Baird, D., and H. Milne. 1981. "Energy Flow in the Ythan Estuary, Aberdeenshire, Scotland." *Estuarine and Coastal Shelf Science* 13:455–472.

Barabási, A. L. 2002. *Linked: The New Science of Networks.* Cambridge: Perseus.

Barabási, A. L., and R. Albert. 1999. "Emergence of Scaling in Random Networks." *Science* 286:509–512.

Barabási, A. L., E. Ravasz, and T. Vicsek. 2001. "Deterministic Scale-Free Networks." *Physica* A 299:559–564.

Berlow, E. L. 1999. "Strong Effects of Weak Interactions in Ecological Communities. *Nature* 398:330–334.

Berlow, E. L., S. A. Navarrete, C. J. Briggs, M. E. Power, and B. A. Menge. 1999. "Quantifying Variation in the Strengths of Species Interactions." *Ecology* 80:2206–2224.

Bodin, Ö., and J. Norberg. 2005. "Information Network Topologies for Enhanced Local Adaptive Management." *Environmental Management* 35:175–193.

Bodin, Ö., M. Tengö, A. Norman, J. Lundberg, and T. Elmqvist. 2006. "The Value of Small Size: Loss of Forest Patches and Threshold Effects on Ecosystem Services in Southern Madagascar." *Ecological Applications* 16:440–451.

Borer, E. T., K. Anderson, C. A. Blanchette, B. Broitman, S. D. Cooper, and B. S. Halpern. 2002. Topological Approaches to Food Web Analyses: A Few Modifications May Improve Our Insights." *Oikos* 99: 397–401.

Borgatti, S. P., M. G. Everett, and P. R. Shirey. 1990. "LS Sets, Lambda Sets and Other Cohesive Subsets." *Social Networks* 12:337–357.

Borrvall, C., B. Ebenman, and T. Jonsson. 2000. "Biodiversity Lessens the Risk of Cascading Extinction in Model Food Webs." *Ecology Letters* 3: 131–136.

Breiger, R., K. Carley, and P. Pattison, eds. 2003. *Dynamic Social Network Modeling and Analysis: Workshop Summary and Papers*. Washington, DC: National Academies.

Brewer, D. D., and C. M. Webster. 1999. Forgetting of Friends and Its Effects on Measuring Friendship Networks." *Social Networks* 21:361–373.

Brooks, C. P. 2006. Quantifying Population Substructure: Extending the Graph-Theoretic Approach." *Ecology* 87:864–872.

Brose, U., R. J. Williams, and N. D. Martinez. 2003. "Comment on 'Foraging Adaptation and the Relationship Between Food-Web Complexity and Stability.'" *Science* 301:918b.

Bunn, A. G., D. L. Urban, and T. H. Keitt. 2000. "Landscape Connectivity: A Conservation Application of Graph Theory." *Journal of Environmental Management* 59:265–278.

Burt, R. 1992. *Structural Holes: The Social Structure of Competition*. Cambridge: Harvard University Press.

Carley, K., and D. Krackhardt. 2001. *A Typology for Network Measures for Organizations*. In CASOS Working Papers. http://www.casos.cs.cmu.edu/publications/papers/org-general.pdf.

Carlsson, L., and F. Berkes. 2005. "Co-management: Concepts and Methodological Implications." *Journal of Environmental Management* 75:65–76.

Cook, K. S., R. M. Emerson, and M. R. Gillmore. 1983. "The Distribution of Power in Exchange Networks: Theory and Experimental Results." *American Journal of Sociology* 89:275–305.

Degenne, A., and M. Forsé. 1999. *Introducing Social Networks*. London: Sage.

den Boer, P. J. 1968. "Spreading of Risk and Stabilization of Animal Numbers." *Acta Biotheoretica* 18:165–194.

Dodds, P. S., D. J. Watts, and C. F. Sabel. 2003. "Information Exchange and the Robustness of Organizational Networks." *Proceedings of the National Academy of Sciences* 100:12516–12521.

Doreian, P., and K. L. Woodard. 1992. "Fixed List Versus Snowball Selection of Social Networks." *Social Science Research* 21:216–233.

Dunne, J. A., R. J. Williams, and N. D. Martinez. 2002a. "Network Structure and Biodiversity Loss in Food Webs: Robustness Increases with Connectance." *Ecological Letters* 5:558–567.

Dunne, J. A., R. J. Williams, and N. D. Martinez. 2002b. "Food-Web Structure and Network Theory: The Role of Connectance and Size." *Proceedings of the National Academy of Sciences* 99:12917–12922.

Ebenman, B., R. Law, and C. Borrvall. 2004. "Community Viability Analysis: The Response of Ecological Communities to Species Loss." *Ecology* 85:2591–2600.

Ehrlich, P., and B. Walker. 1998. "Rivets and Redundancy." *Bioscience* 48: 387.

Elton, C. 1927. *Animal Ecology.* London: Sidgwick and Jackson.

Estes, J. A., and D. O. Duggins. 1995. "Sea Otters and Kelp Forests in Alaska-Generality and Variation in a Community Ecological Paradigm." *Ecological Monographs* 65:75–100.

Estes, J. A., and J. F. Palmisano. 1974. "Sea Otters: Their Role in Structuring Nearshore Communities." *Science* 185:1058–1060.

Everett, M. 1998. *Cohesive Subgroups.* http://www.analytictech.com/networks/chapter_4_cohesive_subgroups.htm.

Fagan, W. F., and L. E. Hurd. 1994. "Hatch Density Variation of a Generalist Arthropod Predator-Population Consequences and Community Impact." *Ecology* 85:2022–2032.

Folke, C., J. Colding, and F. Berkes. 2003. "Synthesis: Building Resilience and Adaptive Capacity in Social-Ecological Systems." In F. Berkes, C. Folke, and J. Colding, eds., *Navigating Social-Ecological Systems: Building Resilience for Complexity and Change*, pp. 352–387. Cambridge: Cambridge University Press.

Fortuna, M. A., and J. Bascompte. 2006. "Habitat Loss and the Structure of Plant-Animal Mutualistic Networks." *Ecology Letters* 9:281–286.

Frank, O. 1981. "A Survey of Statistical Methods for Graph Analysis." In S. Leinhardt, ed., *Sociological Methodology*, pp. 110–155. San Francisco: Jossey-Bass.

Freeman, L. 1979. "Centrality in Social Networks: Conceptual Clarifications." *Social Networks* 1:215–239.

Freeman, L. 2004. *The Development of Social Network Analysis: A Study in the Sociology of Science.* Vancouver: Empirical.

Gadgil, M., P. R. S. Rao, G. Utkarsh, P. Pramod, and A. Chhatre. 2000. "New Meanings for Old Knowledge: The People's Biodiversity Registers Program." *Ecological Applications* 10:1307–1317.

Garlaschelli, D., G. Calarelli, and L. Pietronero. 2003. "Universal Scaling Relations in Food Webs." *Nature* 423:165–168.

Gladwell, M. 2002. *The Tipping Point: How Little Things Can Make a Big Difference.* Boston: Little, Brown.

Goldwasser, L., and J. Roughgarden. 1993. "Construction and Analysis of a Large Caribbean Food Web." *Ecology* 74:1216–1233.

Goodman, L.A. 1961. "Snowball Sampling." *Annuals of Mathematical Statistics* 32:148–170.

Gould, R.V., and R.M. Fernandez. 1989. "Structure of Mediation: A Formal Approach to Brokerage in Transactions Networks." *Sociological Methodology* 19:89–126.

Granovetter, M. 1973. "The Strength of Weak Ties." *American Journal of Sociology* 76:1360–1380.

Gunderson, L.H., and C.S. Holling. 2002. *Panarchy: Understanding Transformations in Human and Natural Systems.* Washington, DC: Island.

Hilborn, R. 1992. "Can Fisheries Agencies Learn from Experience." *Fisheries* 17:6–14.

Holling, C.S. 1973. "Resilience and Stability of Ecological Systems." *Annual Review of Ecology and Systematics* 4:1–23.

Holme, P. 2002. "Edge Overload Breakdown in Evolving Networks." *Physical Review E,* vol. 66.

Holme, P. 2003. "Network Dynamics of Ongoing Social Relationships." *Europhysics Letters* 64:427–433.

Janssen, M.A., Ö. Bodin, J.M. Anderies, T. Elmqvist, H. Ernstson, R.R. McAllister, P. Olsson, and P. Ryan. 2006. "A N etwork Perspective on the Resilience of Social-Ecological Systems." *Ecology and Society* vol. 11, no. 1. http://www.ecologyandsociety.org/vol11/iss1/art15/.

Jen, E. 2005. "Stable or Robust? What's the Difference?" In E. Jen, ed., *Robust Design: A Repertoire of Biological, Ecological, and Engineering Case Studies,* pp. 7–20. New York: Oxford University Press.

Keitt, T.H., A. Franklin, and D.L. Urban. 1995. "Recovery Plan for the Mexican Spotted Owl." In *Landscape Analysis and Metapopulation Structure.* Vol. 2: *Technical and Supporting Information.* Albuquerque, New Mexico: U.S. Department of the Interior Fish and Wildlife Service.

Keitt, T.H., D.L. Urban, and B.T. Milne. 1997. "Detecting Critical Scales in Fragmented Landscapes." *Conservation Ecology,* vol. 1. http://www.consecol.org/vol1/iss1/art4.

Kondoh, M. 2003a. "Foraging Adaptation and the Relationship Between Food-Web Complexity and Stability." *Science* 299:1388–1391.

Kondoh, M. 2003b. "Response to Comment on 'Foraging Adaptation and the Relationship Between Food-Web Complexity and Stability.'" *Science* 301:918c.

Krackhardt, D. 1987. "Cognitive Social Structures." *Social Networks* 9:109–134.

Krause, A.E., K.A. Frank, D.M. Mason, R.E. Ulanowicz, W.W. Taylor. 2003. "Compartments Revealed in Food-Web Structure." *Nature* 426:282–285.

Levin, S. A. 1998. "Ecosystems and the Biosphere as Complex Adaptive Systems." *Ecosystems* 1:431–436.

MacArthur, R. H. 1955. "Fluctuations of Animal Populations and a Measure of Community Stability." *Ecology* 36:533–536.

McCann, K., A. Hastings, and G. R. Huxel. 1998. "Weak Trophic Interactions and the Balance of Nature." *Nature* 395:794–798.

McIntosh, R. J. 2000. "Social Memory in Mande." In R. J. McIntosh, J. A. Tainter, and S. K. McIntosh, eds., *The Way the Wind Blows: Climate, History, and Human Action*, pp. 141–180. New York: Columbia University Press.

Marsden, P. V. 1990. "Network Data and Measurement." *Annual Review of Sociology* 16:435–463.

May, R. M. 1972. Will a Large Complex System Be Stable?" *Nature* 238:413–414.

May, R. M. 1973. "Stability and Complexity in Model Ecosystems." Princeton: Princeton University Press.

Melian, C. J. and J. Bascompte. 2004. "Food Web Cohesion." *Ecology* 82:352–358.

Meyers, L. A., M. E. J. Newman, M. Martin, and S. Schrag. 2003. "Applying Network Theory to Epidemics: Control Measures for *Mycoplasma pneumoniae* Outbreaks." *Emerging Infectious Diseases* 9:204–210.

Milgram, S. 1967. "The Small World Problem." *Psychology Today* 1:61–67.

Milne, H. A., and G. M. Dunnet. 1972. "Standing Crop, Productivity, and Trophic Relations of the Fauna of the Ythan Estuary." In R. S. K. Barnes and J. Green, eds., *The Estuarine Environment*, pp. 86–106. London: Applied Science.

Monasson, R. 1999. "Diffusion, Localization, and Dispersion Relations on 'Small-World' Lattices." *European Physical Journal, B* 12:555–567.

Newman, M. E. J. 2003. "The Structure and Function of Complex Networks." *SIAM Review* 45:167–256.

Newman, M. E. J., and D. J. Watts. 1999. "Renormalization Group Analysis of the Small-World Network Model." *Physics Letters, A* 263:341–346.

Olsson, P. 2003. "Building Capacity for Resilience in Social-Ecological Systems." Ph.D. diss., University of Stockholm.

Olsson, P., and C. Folke. 2001. "Local Ecological Knowledge and Institutional Dynamics for Ecosystem Management: A Study of Lake Racken Watershed, Sweden." *Ecosystems* 4:85–104.

Ostrom, E. 1990. *Governing the Commons: The Evolution of Institutions for Collective Action.* Cambridge: Cambridge University Press.

Ostrom, E. 1998. "Scales, Polycentricity, and Incentives: Designing Complexity to Govern Complexity." In L. D. Guruswarmy and J. A. McNeely, eds., *Protection of Global Diversity: Converging Strategies*, pp. 149–167. Durham: Duke University Press.

Paine, R. T. 1969. "The *Pisaster-Tegula* Interaction: Prey Patches, Predator Food Preference and Inter-tidal Community Structure." *Ecology* 50:950–961.

Paine, R. T. 1980. "Food Webs: Linkage, Interaction Strength and Community In-frastructure." *Journal of Animal Ecology* 49:667–685.

Paine, R. T. 1992. "Food-Web Analysis Through Field Measurement of Per Capita Interaction Strength." *Nature* 355:73–75.

Palla, G., I. Derényi, I. Farkas, and T. Vicsek. 2005. "Uncovering the Overlapping Community Structure of Complex Networks in Nature and Society." *Science* 435:814–818.

Pimm, S. L. 1979a. "The Structure of Food Webs." *Theoretical Population Biology* 16:144–158.

Pimm, S. L. 1979b. "Complexity and Stability—Another Look at MacArthur's Original Hypothesis." *Oikos* 33:351–357.

Pimm, S. L. 1980. "Food Web Design and the Effect of Species Deletion." *Oikos* 35:139–149.

Power, M. E., D. Tilman, J. A. Estes, B. A. Menge, W. J. Bond, L. S. Mills, G. Daily, J. C. Castilla, J. Lubchenco, and R. T. Paine. 1996. "Challenges in the Quest for Keystones." *BioScience* 46:609–620.

Raffaelli, D. G., and S. J. Hall. 1996. "Assessing the Relative Importance of Trophic Links in Food Webs." In G. A. Polis and K. O. Winemiller, eds., *Food Webs: Integration of Patterns and Dynamics*, pp. 165–191. New York: Chapman and Hall.

Redman, C. L., and A. P. Kinzig. 2003. "Resilience of Past Landscapes: Resilience Theory, Society, and the Longue Durée." *Conservation Ecology*, vol. 7.

Rozdilsky, I. D., L. Stone, and A. Solow. 2004. "The Effects of Interaction Com-partments on Stability for Competitive Systems." *Journal of Theoretical Biology* 227:277–282.

Sabidussi, G. 1966. "The Centrality Index of a Graph." *Psychometrika* 31:581–603.

Schneider, M., J. Scholz, M. Lubell, D. Mindruta, and M. Edwardsen. 2003. "Building Consensual Institutions: Networks and the National Estuary Pro-gram." *American Journal of Political Science* 47:143–158.

Seidman, S. B., and B. L. Foster. 1978. "A Graph-Theoretical Generalization of the Clique Concept." *Journal of Mathematical Sociology* 6:139–154.

Shargel, B., H. Sayama, I. R. Epstein, and Y. Bar-Yam. 2003. "Optimization of Ro-bustness and Connectivity in Complex Networks." *Physical Review Letters* 90: art. no. -068701.

Simon, H. A. 1962. "The Architecture of Complexity." *Proceedings of the American Philosophical Society* 106:467–482.

Solé, R. V., and J. M. Montoya. 2001. "Complexity and Fragility in Ecological Net-works." *Proceedings of the Royal Society of London, Series B* 268: 2039–2045.

Solow, A. R., C. Costello, and A. Beet. 1999. "On an Early Result on Stability and Complexity." *American Naturalist* 154:587–588.

Soule, M. E., J. A. Estes, J. Berger, and C. Marinez del Rio. 2003. "Ecological Effectiveness: Conservation Goals for Interactive Species." *Conservation Biology* 17:1238–1250.

Steneck, R. S., Graham, M. H. , Bourque, B. J., Corbett, D., Erlandson, J. M, Estes, J. A., and Tegner, M. J. 2002. "Kelp Forest Ecosystem: Biodiversity, Stability, Resilience and Their Future." *Environmental Conservation* 29:436–459.

Teng, J. and K. S. McCann. 2004. "Dynamics of Compartmented and Reticulate Food Webs in Relation to Energetic Flows." *American Naturalist* 164: 85–100.

Urban, D. and T. Keitt. 2001. "Landscape Connectivity: A Graph-Theoretic Perspective." *Ecology* 82:1205–1218.

van der Leeuw, S. E. 2000. "Land Degradation as a Socionatural Process," in J. A. McIntosh, J. A. Tainter, and S. K. McIntosh, eds., *The Way the Wind Blows.* New York: Columbia University Press.

Wasserman, S., and K. Faust. 1994. *Social Network Analysis: Methods and Applications.* Cambridge: Cambridge University Press.

Watts, D. J. 1998. *Small Worlds: The Dynamics of Networks Between Order and Randomness.* Princeton: Princeton University Press.

Watts, D. J., and S. H. Strogatz. 1998. "Collective Dynamics of 'Small-World' Networks." *Nature* 393:440–442.

Webb, C. T. 2008. "The Role of Spatial Modularity in Ecosystem Resilience." Unpublished MS.

Webb, C. T., and S. A. Levin. 2005. "Cross-system Perspectives on the Ecology and Evolution of Resilience." In E. Jen, ed., *Robust Design: A Repertoire of Biological, Ecological, and Engineering Case Studies*, pp. 151–172. New York: Oxford University Press.

Williams, R. J., E. L. Berlow, J. A. Dunne, A. Barbasi, and N. D. Martinez. 2002. "Two Degrees of Separation in Complex Food Webs." *Proceedings of the National Academy of Sciences* 99:12913–12916.

Woolhouse, M. E. J., C. Dye, J. F. Etard, T. Smith, J. D. Charlwood, G. P. Garnett, P. Hagan, J. L. K. Hii, P. D. Ndhlovu, R. J. Quinnell, C. H. Watts, C. K. Chandiwana, and R. M. Anderson. 1997. "Heterogeneities in the Transmission of Infectious Agents: Implications for the Design of Control Programs." *Proceedings of the National Academy of Sciences, USA* 94: 338–342.

Wootton, J. T. 1997. "Estimates and Tests of Per Capita Interaction Strength: Diet, Abundance, and Impact of Intertidally Foraging Birds." *Ecological Monographs* 67:45–64.

4

SOCIAL NETWORKS AS SOURCES OF RESILIENCE IN SOCIAL-ECOLOGICAL SYSTEMS

Thomas Hahn, Lisen Schultz, Carl Folke, and Per Olsson

GOVERNANCE AND MANAGEMENT

GOVERNANCE OF ECOSYSTEMS or social-ecological systems has lately received increasing attention (Dietz, Ostrom, and Stern 2003; Eckerberg and Joas 2004; Folke et al. 2005; Ostrom 2005). If management is about strategies for handling natural resources, governance addresses the broader social contexts of creating the conditions for social coordination that enable ecosystem-based management (Stoker 1998; Lee 2003). Boyle, Kay, and Pond (2001) have suggested a triad of activities where governance is the process of resolving trade-offs and providing a vision and direction for sustainability, management is the operationalization of this vision, and monitoring provides feedback and synthesizes the observations to a narrative of how the situation has emerged and might unfold in the future.

Social networks have been shown to play a crucial role in each of these three activities. For instance, Scheffer, Westley, and Brock (2003) have suggested that a clear and convincing vision, comprehensive stories, and meaning as well as good social links and trust with fellow stakeholders may mobilize several interest groups at several organizational levels and start a self-organizing process of learning and social capital generation for management of complex adaptive ecosystem.

In this chapter we illuminate, using one case study from southern Sweden, the crucial role of multilevel social networks for generating visions and ecological knowledge and connecting this to management and governance of a social-ecological system. The social networks of Kristianstads Vattenrike appear to have succeeded in transforming a social-ecological system toward a more sustainable trajectory and building resilience in this new trajectory.

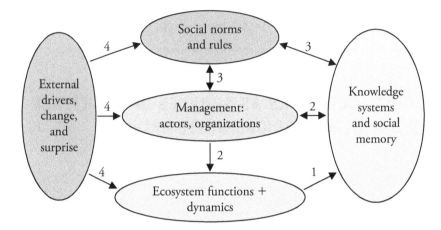

FIGURE 4.1. A conceptual model of the dynamics facing a linked social-ecological system (SES). An SES consists of an ecosystem, the management of this ecosystem by actors and organizations, and the formal and informal institutions (rules, social norms, and conventions) underlying this management. The resilience of an SES depends on ecological dynamics as well as the organizational and institutional capacity to adapt to eco- system dynamics. This requires a learning environment and links between key persons across organizational levels. To be resilient, the social-ecological system also needs capacity for dealing with external change. (Modified from Hahn et al. 2006.)

We discuss how learning and trust building thrive in informal collabora- tion, which has also been noted by Westley (1995) and Gunderson (1999). We also show how conflicts have been resolved on a deeper level by empha- sizing knowledge generation and worldviews, i.e., perceptions and assump- tions on the relationships between nature, humans, and society (Adams et al. 2003).

In a recent review (Folke et al. 2005) we concluded that successful adap- tive approaches for ecosystem management under uncertainty involve (figure 4.1)

1. *building knowledge and understanding of resource and ecosystem dynam- ics*: detecting and responding to environmental feedback in a fashion that contributes to resilience require ecological knowledge and understanding of ecosystem processes and functions. Hence managers need to mobilize all sources of understanding to reduce ecological illiteracy. This involves linking people and organizations with different knowledge systems (Gadgil, Olsson et al. 2003).

2. *feeding ecological knowledge into adaptive management practices*: success- ful management is characterized by continuous testing, monitoring, and

reevaluation to enhance adaptive responses acknowledging the inherent uncertainty in complex systems. Management needs to adapt to new knowledge and build this into management plans rather than striving for optimization based on past records (Berkes, Folke, and Colding 2003). Forming a learning environment that accepts continuous testing and changes requires leadership within management organizations (Danter et al. 2000) and collaboration within social networks.

3. *supporting flexible institutions and multilevel governance systems*: the adaptive governance framework is operationalized through adaptive comanagement (Olsson, Folke, and Berkes 2004) where the dynamic learning characteristic of adaptive management is combined with the multilevel linkage characteristic of comanagement (Pinkerton 1989). The sharing of management power and responsibility may involve multiple, often polycentric, cross-level institutional and organizational linkages among user groups or communities, government agencies, and nongovernmental organizations, i.e., neither centralization nor decentralization (Ostrom 1998). This collaboration and adaptive governance draws on visions and narratives from the social memory of past ecological crises and responses and requires enabling legislation and social incentives for collaboration (Hahn et al. 2006). Social networks are instrumental for mobilizing social memory, generating social capital as well as legal, political, and financial support to ecosystem management initiatives.

4. *dealing with external drivers, change and surprise*: it is not sufficient for a well-functioning multilevel governance system to be in tune with the dynamics of the ecosystems under management (referred to as the "internal resilience"; by Folke et al. 2004). It also needs to develop a capacity for dealing with changes in climate, disease outbreaks, hurricanes, global market demands, subsidies, and governmental policies. The challenge for the social-ecological system is to accept uncertainty, be prepared for change and surprise, and enhance the adaptive capacity to deal with disturbance. Nonresilient social-ecological systems are vulnerable to external change, while a resilient system may even make use of disturbances as opportunities to transform into more desired states.

Adaptive governance of ecosystems involves a large number of key persons with different skills who perform different leadership functions. Borrowing a metaphor from ecology,[1] Folke, Colding, and Berkes (2003) suggest that successful ecosystem management systems involve different actor groups who perform some critical functions. In Kristianstads Vattenrike we have identified knowledge carriers, knowledge generators, stewards, leaders, and sense makers (Olsson, Folke, and Hahn 2004). Based on several other case studies, Folke, Colding, and Berkes (2003) also identified the following key groups; knowledge retainers, interpreters, facilitators, visionaries, inspirers, innovators, experimenters, followers, and reinforcers. Some key persons perform several functions. One indispensable role is to build trust, invest in personal relations, and develop networks across stakeholder groups. This enhances the social capital and facilitates conflict resolution. The absence of an actor group performing a particular key function or extreme reliance on one or a few key persons may indicate vulnerability of an ecosystem management system.

In this chapter we argue that the adaptive capacity of a social-ecological system is related to the existence of social networks. These often emerge as self-organizing processes (i.e., not guided by external pressure) involving key persons who share some common interests although they represent different stakeholder groups (McCay 2002). Organizations that do not appear to have much in common may develop crucial links thanks to these key persons who form the nodes of different, loosely connected, networks.

The size of a network can be viewed as the number of nodes and links or ties. The strength of a network depends on the ability of the nodes, the key persons, to exchange information with other stakeholders, identify common interests, and gather support for such interests (e.g., ecosystem management) within their own organization or stakeholder group. In his seminal paper Granovetter (1973) argued that weak ties, i.e., the bridges *between* different stakeholder groups, may be the most valuable for generating new knowledge and identifying new opportunities and hence create a macro effect: "those to whom we are weakly tied are more likely to move in circles different from our own and will thus have access to information different from that which we receive" (Granovetter 1973:1371). Applied to ecosystem management, we argue that a loosely connected network involving a diversity of stakeholders is important for gathering different types of ecological knowledge, building moral and political support (legitimacy) from "nonenvironmental" sectors, and attaining legal and financial support from various institutions and organizations. Hence, if polycentric cross-level institutions provide the structure for adaptive comanagement, multiple overlapping networks of key persons provide the processes.

We focus on sources of resilience and stress the role of social capital (Westley 1995; Adger 2003) for building social networks for ecosystem management. Pretty and Ward (2001) refer to social capital as relations of trust, reciprocity, common rules, norms and sanctions, and connectedness in institutions. In the common property literature the focus has been on the strong ties (bonding) within a community of resource users (Ostrom et al. 1999) while for governance of social-ecological systems (SESs) the challenge is to invest in social capital in the weak ties to bridge the interests of different stakeholder groups (Adger and Tompkins 2004; Newman and Dale 2005; Olsson et al. 2006). These relationships must be fed with relevant visions and knowledge on ecosystem dynamics. A special kind of knowledge is the social memory (McIntosh 2000) for dealing with uncertainty and external change and disturbance across ecological scales and organizational levels.

To illuminate the role of social networks as sources of resilience we synthesize findings from our studies of Kristianstads Vattenrike (KV) in southern Sweden, where we have focused on understanding the social dimension of the emergence of an ecosystem-based approach to management of a wetland landscape in a semiurban area. We describe how social networks were formed for three different purposes—generating ecological knowledge, horizontal collaboration for legitimizing ecosystem management, and vertical collaboration for gaining financial and political support through multilevel policy communities.

We describe how key persons formed these networks, drawing on social memory and social capital, and transformed the social-ecological system into a new trajectory. We highlight the dynamic social processes involved in deepening and widening the new stability landscape—the social sources of resilience in social-ecological systems. In particular, we emphasize the role of leadership, vision, sense making, and the dynamics of adaptive governance, that is, how local actors develop a dynamic policy community (Shannon 1998) to navigate the institutional and organizational structures. This synthesis highlights the significance of what we refer to as bridging organizations for building trust among stakeholders, mobilizing knowledge, solving conflicts, and linking governance at several organizational levels (Folke et al. 2005).

THE SOCIAL NETWORKS OF KRISTIANSTADS VATTENRIKE

Kristianstads Vattenrike, literally meaning the "water riches or water realm of Kristianstad," was proposed in 1989 to describe an area of 1,100 sq km defined by hydrological and political borders. It includes the Helgeå River catchments and the coastal regions of Hanö Bay within the Municipality of Kristianstad. In June 2005 KV eventually became the first Man and Biosphere (MAB) reserve in Sweden to fulfill the new Sevilla criteria. The lowland area has long been appreciated for its cultural and natural values. Besides high biodiversity, unique habitats, and aesthetic landscapes, it is one of Sweden's most productive agricultural areas. The abundance of ecosystem services generated in the area is also reflected in the diversity of steward organizations, from local farmers to international nature conservation organizations (Olsson, Folke, and Hahn 2004; Schultz et al. 2004).

The idea of KV came from a small group of concerned local inhabitants. Their proposal was well received by the Municipality Board, which established Ecomuseum Kristianstads Vattenrike Biosphere Reserve (EKV), a small municipal organization with the aim of enhancing the ecosystem services and values of the area of Kristianstads Vattenrike.

From the KV case it is possible to distinguish three networks with different purposes and functions: the first concerned with management and generating local ecological knowledge, the second with coordinating multiple objectives within the municipality, and the third with multilevel policy communities. In KV these three networks emerged simultaneously and continue to grow today, eighteen years after it all started. The experiences from KV concerning each of these network functions are discussed below.

NETWORKS FOR MOBILIZING LOCAL STEWARDS AND GENERATING LOCAL ECOLOGICAL KNOWLEDGE

Baseline information about ecosystem functions and performance is needed for identifying and quantifying different ecosystem services as well as managing the ecosystems generating these services.[2] For wetlands and other ecosystem types there are general estimates of which ecosystem services are typically produced, but these values need to be assessed and communicated in each location in order to be appreciated by people within a particular social-ecological system. Site-specific ecosystem knowledge is often generated and held by people involved in local management projects (Olsson and Folke 2001).

In the early 1980s there were several projects concerned with ecosystem and environmental management in Kristianstad. Some were initiated by the Municipality of Kristianstad and involved actors at multiple organizational levels. However, the existing projects were generally narrow in scope and their participants were often unaware of the other projects (Olsson et al. 2007). Several ecosystem services were declining, in spite of the fact that the core area of what is now known as Kristianstads Vattenrike was already acknowledged as a wetland area of international importance by the Ramsar Convention on Wetlands in 1975 (Olsson, Folke, and Hahn 2004). A decline in bird populations, dependent on cultivated flooded meadows, was reported in 1979 by the local bird-watching society. One of the key persons in forming network A was a member of the local bird-watching society and was also working at the County Administrative Board (CAB). Here we call this person HC.

At this time another key person, SEM, who became the director of EKV in 1989, was working at the County Museum. For the two-hundred-year anniversary of Linneus's death in 1978, he had made exhibitions on Linneus's narratives from his trips in the area of Kristianstad; since then he had synthesized existing information combining ecological and historical maps. The area was found to have unique values thanks to the vast flooded meadows that had been used for grazing and/or hay making for several hundred years. Most other flooded meadows in Sweden had been abandoned and overgrown by willow bushes and wet forests. Linking to the CAB and the local bird-watching society, he coordinated on-ground inventories of the wetlands. These mappings and inventories finally resulted in articulation and quantification of unique values in 1989, forming the foundation of KV. In this process, coarse-grained scientific information from national surveys were used to guide the local efforts to produce fine-grained, context-specific knowledge needed for developing detailed management plans (Olsson, Folke, and Hahn 2004). A third key person, MD, who was an environmental manager at the Technical Municipal Administration responsible for halting eutrophication, was also involved in this network for synthesizing and generating ecological knowledge (figure 4.2).

Since 1989 EKV and the CAB have adopted a collaborative approach toward farmers and local steward organizations. Local ecological knowledge has been utilized and expanded through a collaborative learning process involving a lot of small inventories of various species at individual farms or tributaries to the Helgeå River. Sometimes these were initiatives by EKV, sometimes by the bird-watching society, the society for nature conservation, or a fishing association. Through habitat restoration and new management,

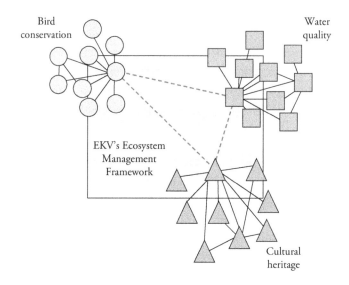

FIGURE 4.2. Before 1989 three independent clusters of local people were concerned about the decline in water quality, bird populations and the ceased cultivation of flooded meadows with high biological and cultural heritage values. These declining trends were correlated, but the three networks were single focused. In 1989 SEM, the employee at the County Museum, synthesized ecological and historical knowledge and invested in personal links with one key person from each of the other two networks: HC and MD. The dotted lines in this sketch indicate these new connections between nodes of different network (Olsson et al. 2007).

several species have recovered, and some have been reintroduced. As a result, EKV received the Conservation Award 2002 from the Swedish Species Information Center (ArtDatabanken) for its "systematic work on integrating the values of the wetlands into the ordinary operations of the municipality."[3]

The three initial clusters for knowledge generation have been expanded dramatically since 1989. Today, twenty to thirty local steward organizations involving several hundred people—farmers, tourist entrepreneurs, as well as members of associations for hunting, fishing, bird-watching, conservation, and village cultural heritage (*byalag*)—are involved in a loosely connected local network that generates ecological knowledge (Schultz, Olsson et al. 2004). A fourth key person, here called KM, is in charge of EKV's information and communication, including the process of becoming a MAB reserve. She has also developed the Nature School, which facilitates teachers and students actively monitoring ecosystem change in the landscape. All parts of the network are not active all the time, but knowledge and memory of ecosystem dynamics that resides among various stewards can be mobilized for particular projects. We will return to the project organization of EKV in the last section of the chapter.

One example of a network oriented for ecological knowledge generation and conservation is the Frog Project (figure 4.3). This project was initiated by the local branch of a Swedish NGO, the Swedish Society for Nature Conservation, after a call for "frog watchers" from an external expert, employed at the National Agency for Conservation and researcher at Gothenburg University. Frog watchers are volunteers involved in monitoring, caretaking, and protection of habitats and populations of threatened amphibian species. The frog group was formed in Kristianstad, engaging around twenty volunteers in mapping existing amphibian populations, improving habitats, and continuous monitoring. As described earlier, coarse-grained information from larger surveys guided the local inventories. The network around the frog group provides financing, permissions for constructing habitats, as well as opportunities for learning and exchanging information. The group

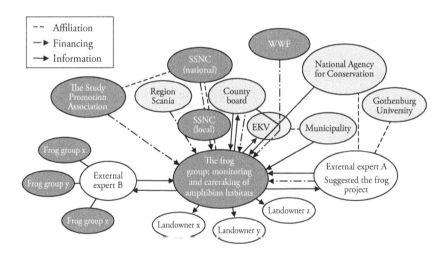

FIGURE 4.3. A network formed around conservation and monitoring of amphibians, mainly Bufo calamita, in Kristianstads Vattenrike. The group was formed within the local branch of the SSNC (Swedish Society for Nature Conservation) after a call for "frog watchers" from an external expert, employed at the National Agency for Conservation and doing research at Gothenburg University. The group conducts fine-grained inventories of the threatened species and notifies the county board in case of decline or harmful activities. Dark circles represent NGOs, gray circles represent governmental organizations, and white circles are individuals. Information arrows pointing to the frog group represent earlier inventories, biological information, maps, and information on inventory methods. Arrows originating in the frog group represent information on status of amphibian habitats and populations. "Affiliation" lines connect groups belonging to the same umbrella organization (Schultz, Folke, and Olsson 2007).

is mainly concerned with the red-listed species *Bufo calamita* and reports declines in populations as well as threatening activities such as filling of habitats to the county board. Several times it has succeeded in protecting and improving habitats important to the local populations. Information generated by the frog group is also used in research projects and spread to landowners with important habitats on their properties.

NETWORKS FOR COORDINATING MULTIPLE LOCAL OBJECTIVES AND SEIZING A WINDOW OF OPPORTUNITY

As previously described, the links between three key persons (see figure 4.2) initiated the process that resulted in KV and EKV. However, before sending the proposal about KV to the Municipality Board, SEM (the EKV director to be) contacted representatives from other sectors and developed a broad vision similar to what is now known as *the ecosystem approach* adopted by the Convention on Biological Diversity[4] and the UNESCO Man and Biosphere reserve.[5] Indeed, the idea to make KV a MAB reserve was already present in the original proposal (Olsson, Folke, and Hahn 2004), which illustrates that a firm vision and direction can be combined with a flexible project organization (Hahn et al. 2006).

The land use mapping and inventories discussed before were essential to quantify cultural heritage and ecological values. The result from this monitoring was synthesized into a direction for change in the form of overall goals and a vision of a holistic approach to wetland management or, as Boyle, Kay, and Pond (2001) put it, into a narrative of how the system has emerged and might unfold in the future. The vision was (and is) to preserve and develop the ecological values and cultural heritage of the area, while at the same time making careful and sustainable use of these values, turning the wetland into a water realm that would put Kristianstad on the map. In this process SEM coined the concept Kristianstads Vattenrike, and, by suggesting how footpaths and outdoor museums would make the "swamp" more accessible, he managed to change the perception of the area from water sick to water rich. The challenge was double: increasing the legitimacy of an ecosystem-based management of public and private land and water resources, and tuning this management so that it contributed to other social goals as well. Five key persons were involved in the proposal accepted by the Municipal Board in 1989 (Olsson, Folke, and Hahn 2004):

1. a researcher at Lund University, interested in using wetlands for reducing eutrophication,
2. an official at WWF Sweden, willing to finance nature conservation projects,
3. the rector at Kristianstad University College, interested in opportunities for research and education,
4. a hotel director, previously president of the Kristianstad Tourism Board, interested in ecotourism, and
5. the director of the National Museum of Natural History.

Not all these five key persons were local but the fact that they represented several social sectors impressed the local decision makers of the Municipality Board (Olsson, Folke, and Hahn 2004). In addition, SEM approached representatives from the CAB and the local branches of the Farmers' Federation and the Society for Nature Conservation individually face-to-face to discuss the vision, identify common interests, and build trust for future collaboration. In discussions with persons representing other sectors, he focused on specific parts of the proposal that this particular person and his/her organization would be interested in. For instance, the CAB had responsibilities for regional development and liked the idea of ecosystem-based development. The CAB was prepared to let one employee (the key person HC mentioned earlier) work half-time with issues relevant to both the CAB and to KV, on condition that the municipality also contributed. The Municipality Board, intrigued by the opportunity to put Kristianstad "on the map" (in which they have definitely have succeeded), agreed to establish Ecomuseum Kristianstads Vattenrike (EKV) and employ SEM as the director. And WWF was willing to finance explicit project costs for conservation that neither the municipality nor the CAB was willing to do (Olsson, Folke, and Hahn 2004).

The contact between SEM and a local top politician of the Municipality Board provided a crucial link at a critical time, an aspect that has also been emphasized by Westley (2002). This politician was enthusiastic about the holistic approach and the suggested name Kristianstads Vattenrike. The politician notes that "SEM presented the area in a different way than anyone had done before, and I became aware of the values. Many considered the wetlands as a problem. . . . SEM presented a nature conservancy plan that didn't close the area but opened it up and made it accessible for the public . . . . He managed to engage and involve several important groups in the project, even farmers."

This enabled SEM to take advantage of a window of opportunity (a policy window; Kingdon 1995). No single issue had ever received as much attention during a national election as the environmental issue received in 1988; ecosystem issues were pressing, and this opened the window for a few years at the end of the 1980s. "Had we not taken the chance at that time we would still be knocking on the door today," as SEM expressed it. The other two factors that need to be in place simultaneously for a policy change according to Kingdon (1995)—a policy proposal and a politician who is willing to work it through the political process—were also present. As a result, the window of opportunity could be seized (Olsson, Folke, and Hahn 2004).

Today, there are several multistakeholder projects going on that are coordinated by EKV. The largest of them, measured by the number of people actively involved, are the Flooded Meadows Project (described under C; see figure 4.5) and the Crane Project (Hahn et al. 2006), the Sandy Grasslands Project (Schultz et al. 2004), and the Vramsån Creek Project. The Crane Project illuminates the importance of trust building and early response. Since 1997 the number of cranes passing through Kristianstad has been increasing, resulting in crop damage. As a response to growing discontent among farmers, EKV gathered a small group of "like-minded" farmers and bird-watchers to a meeting to forestall conflict escalation, which, according to SEM, "would probably have stifled learning and eroded trust . . . . Without that meeting the farmers' organization would probably have developed their own policy and strategy only looking at their own interest." A farmer from Hornborgarsjön, the number one bird lake in Sweden, also came to this meeting to share their experiences with the local farmers on how to handle cranes. SEM's motivation: "The best thing is if people who speak the same 'language' share knowledge and enthusiasm. This is a general strategy." Since then the Crane Group is run by farmers who volunteer to assist other farmers with experiments and equipment to minimize crop damage. It has also arranged bird-watching events (Hahn et al. 2006).

The Vramsån Creek Project (figure 4.4) was launched in 1999 as a joint project between WWF and the EKV and involves landowners and potato farmers around the creek as well as different NGOs such as fishing associations and the bird-watching society. The project aims at improving biodiversity by restoring habitats and changing management practices associated with the Vramsån Creek, one of Scania's most valuable creeks in terms of nature conservation.

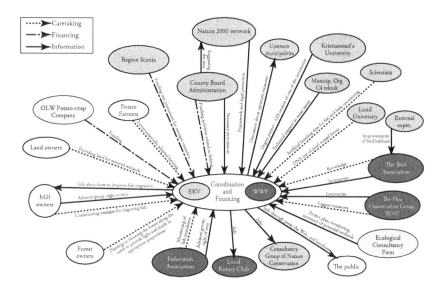

FIGURE 4.4. The Vramsån Creek Project is coordinated and financed by the WWF and EKV and draws upon a large network of local steward associations/NGOs, governmental organizations, and the private sector. The network provides financial and legal support as well as ecological knowledge, monitoring, and caretaking (e.g., improving habitats). Dark circles represent NGOs, gray circles represent governmental organizations, dotted circles represent private companies, and white circles represent individuals acting outside organizations. (Modified from Schultz et al. 2004.)

MULTILEVEL NETWORKS AS A POLICY COMMUNITY

The social arrangements for governing the ecosystems of KV can be referred to as a policy community or policy network. A policy community has been defined as a diverse network of individuals representing public and private organizations generally associated with the formation and implementations of policy in a given resource area (Shannon 1998).

The policy community around EKV had already emerged in 1989 during the process of the first proposal and it has expanded ever since. Being a subset of the whole loosely connected network of persons around EKV, the policy community includes the more policy-inclined actors, mainly those at higher organizational levels but also some of the actors at the local level. From our studies it is clear that the policy community does not consist of various organizations and authorities but key persons representing these organizations and authorities (Hahn et al. 2006). The key persons help EKV to navigate the larger legal, political, and financial environments.

The policy community is framed in local ecosystem contexts; it is formed recognizing site-specific environmental and social conditions and links policy making at the local, district, and national levels. Without this policy community it would be very costly to establish the contacts needed for each individual project to be successful. All of the larger project that are coordinated by EKV include parts of the policy community. The Flooded Meadows Project is by far the largest in monetary terms since it depends heavily upon subsidies from the Common Agricultural Policy of the EU. It is also the project that best illustrates the multilevel policy community around EKV (figure 4.5).

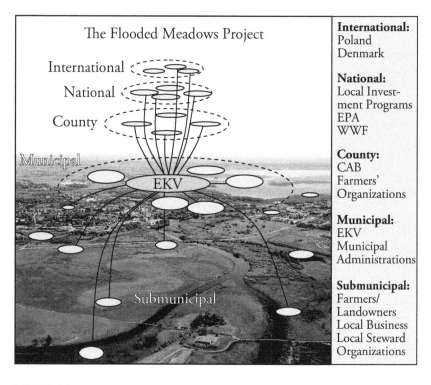

FIGURE 4.5. The urban part of Kristianstads Vattenrike (KV) at summer, showing the wetlands, the two branches of Helgeå River running from Lake Araslövsjön to Lake Hammarsjön in the forefront, and the Hanö Bay of the Baltic Sea in the background (photo Patrik Olofsson). The social network of the Flooded Meadows Project of KV has benefited from experiences of wetland management in Poland and Denmark. Each node includes one or several key persons, often employed by or members of an organization. Ecomuseum Kristianstads Vattenrike (EKV) is the central node of the network. The cross-level collaboration has started as informal contacts by EKV and has sometimes become formalized by contracts and joint ventures. (Modified from Hahn et al. 2006.)

TABLE 4.1. Formal Institutions Linked to the EKV

FORMAL INSTITUTIONS	LEVEL OF INSTITUTION
Man and the Biosphere Reserve (MAB)*	Global (UNESCO)
Restoration grants	Regional (EU)**
Environmental subsidies	Regional (EU)**
Local investment grants	National (Government)
Prohibition of embankment	District (CAB)
Nature reserves	District (CAB)

* MAB is indeed a nonstatutory institution, but, because of the enormous political and moral support it enjoys, it works de facto as a formal institution.
** These are administered by the national EPA.
Hahn et al. 2006.

All collaboration, horizontal (local) as well as vertical (multilevel), that is coordinated by EKV and thus under the municipal umbrella is *voluntary*, which implies that there are no formal rules that forces actors to collaborate. This makes the achievements by EKV vulnerable to internal changes in staff or to declining interest of farmers and other actors. However, experiences from KV suggest that institutional arrangements such as nature reserves and other formal agreements between parties about ecosystem management can emerge from the collaborative processes. In this way, and thanks to the increasing local legitimacy of EKV, some achievements from informal collaboration are formalized and vulnerability is reduced.

The relations between local and vertical links are important for understanding the increased legitimacy of an ad hoc organization like EKV. The policy community around EKV enables it to navigate the larger legal, political, and financial environment (table 4.1). Today they have established trust with staff at the Environmental Protection Agency (EPA) and the WWF, which enables small, flexible financial support when a pressing issue arises. The Common Agricultural Policy (CAP) of the EU offers restoration grants for wetlands and environmental subsidies for grazing flooded meadows, and there are sometimes new governmental programs that fit the activities in KV. Taking advantage of these opportunities requires skills other than those needed for local collaboration. Our analysis suggests that EKV manages both types of skills very well (Hahn et al. 2006).

As mentioned before, the achievements from voluntary collaboration are sensitive and vulnerable to sudden changes. Nature reserves are regarded, by EKV, as the best way to formalize such achievements and make them resilient to threats like declining interest by landowners and other stewards or declines in subsidies. Another way for EKV to build resilience in social-ecological systems is to influence physical planning for land use, which is an important tool for Swedish municipalities. Land use plans may prohibit further embankments of wetlands and point out some areas for future nature reserves, thus providing a prelegal direction for future development. EKV participates in the Consultancy Group for Nature Conservation together with a dozen other interest groups to advise the municipality on land use plans.

Recently, the most important part of the policy community has been institutions and organizations at the national and international levels assisting KV to become a Man and Biosphere (MAB) reserve. Although MAB reserves are nonstatutory entities, this provides moral and political support to the ecosystem approach as it is implemented in KV. Thus it secures the achievements from voluntary collaboration and reduces the vulnerability that is inherent in a project organization such as EKV, relying on a few key persons.

THE ADHOCRACY ORGANIZATION OF EKV AND ITS APPROACH TO COLLABORATION

As mentioned earlier, the three types of networks described perform different purposes or functions, all of which seem to be necessary for adaptive governance of social-ecological systems. We are unable and unwilling to generalize about how these different types of social networks emerge and how they are linked. In the KV case the generation and mobilization of ecological knowledge was an integral part of horizontal and vertical collaboration, and SEM and other key persons were involved in all three networks. Hence the different social networks coevolved, and this fact contributes to the organizational form of EKV.

EKV is a project organization, and its activities can be described by the multiple small and large projects its staff coordinates. The monetary cost of the Municipality of Kristianstad for EKV is very small since HC is paid half-time by the CAB and KM is paid by special grants related to the MAB application process. Most of the money needed for different project comes from external sources or from municipal administrations if the projects serve their objectives as well. Several projects have been suggested and initiated by individual farmers, fishing organizations, or other steward organizations. EKV is by no means coordinator or even involved in all projects related to enhancing ecosystem services that are run by the numerous steward organizations mentioned earlier. But EKV decides which activities or projects are undertaken in the name of Kristianstads Vattenrike.

Governing complex adaptive ecosystems requires adaptive managers supported by flexible organizations (Westley 1995); problem-oriented organization or adhocracy organizations (Mintzberg 1979) have been suggested by Danter et al. (2000) and observed by Imperial (2005) as significant in this context. As a project organization, EKV is an adhocracy; it develops or collects good ideas for small projects and waits for the right persons or opportunities to emerge. When a project idea materializes, an ad hoc project organization is formed by reconnecting "old"—and maybe some "new"—people from different networks. The technical knowledge and social memory of ecosystem functions that the small staff of EKV does not possess is "borrowed" and mobilized from these networks. Being a municipal organization, EKV has an unusually free mandate: to catalyze private activities to enhance the values of the ecosystems, "to do the things that normally fall between the chairs of municipal responsibilities," as the EKV director puts it. EKV has no legal authority to regulate or enforce rules, and this is regarded as an asset by everyone we interviewed: a learning atmosphere has emerged where people feel free to share ideas and knowledge with EKV (Hahn et al. 2006).

However, these ad hoc network structures do not replace the accountability of existing hierarchical bureaucracies, but rather operate within and complement them (Kettl 2000). Since all participation is voluntary, EKV needs to identify win-win situations—with farmers, for example. Conservation initiatives by authorities are traditionally communicated with farmers in a top-down fashion, but the EKV approach starts with building trust and listening to farmers' concerns and knowledge. If common interests are identified, which is only possible among those actors sincerely interested in ecosystem management, this may result in an agreement on a *private* basis—with individual farmers, tourist entrepreneurs, and small steward organizations. EKV provides an arena for collaborative learning and value/preference formation, that is, when actors learn how environmental stewardship and conservation can be beneficial for other interests and see how it has worked out at neighboring sites, they may change their preferences.

The EKV director never calls for large meetings with different stakeholders to discuss controversial issues like nature reserves with farmers. His earlier experience from large unconditional meetings is that the most critical farmers would take this opportunity to voice their discontent. After such positioning it is hard to step back and start looking for common interests: "I mean, you don't gather people if you don't think anything positive will come out of the meeting" said SEM. Big meetings do not create an atmosphere of learning, listening, and conflict resolution. EKV's approach is therefore to initiate processes with a clear direction by "elevating the issues" that people otherwise tend to quarrel about and start identifying win-win situations and concrete projects for collaboration. Indeed, win-win projects can often be identified only after a process of trust building and value formation in which ecological concerns are discussed and the visions and goals are deliberated. This is the reason why the social networks around EKV must grow slowly, face-to-face, starting with the most positive individuals and letting their enthusiasm convince the more hesitant (Hahn et al. 2006). We will soon return to the strategic and democratic implications of this private collaboration.

The social networks around EKV are essential for building resilience in the existing trajectory of ecosystem-based management. Olsson, Folke, and Hahn (2004) distinguished between three sets of processes and strategies that contribute to resilience: developing motivation and values for ecosystem management, directing the local context through adaptive comanagement, and navigating the larger environment (table 4.2).

Of these sets the first two are related to the social networks for generating ecological knowledge and horizontal (local) collaboration, the third to vertical (cross-level) collaboration. Besides the factors already discussed, table 4.2 emphasizes how EKV continuously communicates the success and progress of projects within the networks as well as through local and national media. This maintains the necessary commitment and social norms among the key persons of the networks, makes sense of what ecosystem management is about, increasing legitimacy with the public and thus providing a buffer for external drivers (figure 4.1). Another factor highlighted in table 4.2 is the mobilization of external knowledge, including scientific knowledge as well as experiences from managing flooded meadows, wildlife, ecotourism at other sites, nationally as well as internationally. Lately this has included several visits to MAB reserves all over the world.

TABLE 4.2 Social-Ecological Processes and Strategies That Contribute to Resilience

DEVELOPING MOTIVATION AND VALUES FOR ECOSYSTEM MANAGEMENT

Envisioning the future together with actors

Developing, communicating, and building support for the mission

Identifying and clarifying objectives

Developing personal ties

Establishing a close relationship and trust with key individuals

Fostering dialogue with actors

Providing arenas for trust building among actors

Building trust in times of stability to facilitate conflict resolution

Developing norms to avoid loss of trust among actors

Continuously communicating success and progress of projects

DIRECTING THE LOCAL CONTEXT THROUGH ADAPTIVE COMANAGEMENT

Encouraging and supporting actors to perform monitoring, including inventories

Encouraging and supporting actors to manage ecosystem processes for biodiversity and ecosystem services

Initiating and sustaining social networks of key individuals

Mobilizing individuals of social networks in problem-driven projects

Making sense of and guiding the management process

Synthesizing and mobilizing knowledge for ecosystem management

Providing coordination of project and arenas for collaboration

Encouraging and inspiring actors to voluntary participation

Initiating projects and selecting problems that can be turned into possibilities

Creating public opinion and involving local media

NAVIGATING THE LARGER ENVIRONMENT

Influencing decision makers at higher levels to maintain governance structures that allow for adaptive comanagement of the area

Mobilizing new funding when needed

(*continued*)

TABLE 4.2 *(Continued)*

Mobilizing external knowledge when needed
Exchanging information and collaboration with local steward associations in Sweden and internationally
Collaborating with national and international scientists
Collaborating with national and international nongovernmental organizations
Participating in international institutional frameworks
Supporting diffusion of the values of KV through social networks
Providing a buffer for external drivers
Communicating with national media

Social-ecological processes and examples of strategies used for building resilience after the inception of the EKV. These adaptive processes increase the capacity for dealing with uncertainty and change (from Olsson, Folke, and Hahn 2004).

DISCUSSION

EKV can be regarded as a bridging organization that connects and navigates the interests of different stakeholders ("weak ties") as well as across organizational levels. A bridging organization provides an arena for trust building, sense making, identification of common interests, learning, vertical and/ or horizontal collaboration, and conflict resolution (Folke et al. 2005). As an integral part of adaptive governance of social-ecological systems, bridging organizations provide social incentives by rewarding and creating space for collaboration, value formation, and innovation (Hahn 2006). Westley (1995) has used the same term for describing interorganizational collaboration. Bridging organizations differ from boundary organizations that focus on synthesizing and translating scientific knowledge to make it relevant for policy making (Cash and Moser 2000). A bridging organization for adaptive governance is able to address the factors listed in table 4.2.

The collaboration that EKV initiates is *strategic*; it is conditional on the goals to enhance the values from the ecosystems. This is one clear interpretation and operationalization of sustainable development, assuming a hierarchy of its three pillars; there are boundaries for human activities set by ecology and biophysics and hence social and economic development must be conditional to these. These boundaries or limits suggested by scholars within the ecological and biophysical sciences are of course socially constructed—they don't simply exist "out there." They coevolve with human societies and, as Norgaard (1996) has argued, they are useful as metaphors that provide direction for social and economic development. In the EKV approach, fixed goals (enhancing the ecosystem capacity) are combined with a flexible organization. The ecological goals set by EKV, and adopted by the Municipality of Kristianstad, are not only constraints to social and economic development but provide a framework and direction for this development.

With respect to the democratic implications of "the EKV approach," we need to distinguish between private and public ownership. When initiating collaboration with farmers and fishing associations having private property rights, EKV must make attractive proposals (win-win) to influence management. On public (state or municipal) land and water there are often several stakeholders claiming management authority or at least use rights—conservationists, developers or exploiters, and boating associations. Sometimes these conflicts can be resolved in consensus, i.e., the EKV approach, but sometimes it becomes an issue for the Consultancy Group for Nature Conservation. This consultancy group is the only "political" or collective-choice forum in which EKV participates, and as part of this forum they do not, and cannot, select who participates as they do in the private projects.

A crucial challenge for adaptive governance of social-ecological systems seems to be the mobilization of social memory. Social memory has been defined as the arena in which captured experience with change and successful adaptations, embedded in a deeper level of values, is actualized through community debate and decision-making processes into appropriate strategies for dealing with ongoing change (McIntosh 2000). It is a part of the cultural capital of human society (Berkes and Folke 1992) and of significance in analyses of social capital. A subset of social memory is the accumulation of a diversity of experiences concerning management practices and rules in use at the collective level. Social memory provides a foundation for modification of rules, typically refering to decadal time scales as opposed to months or a year (McIntosh 2000). It is important for linking past experiences with present and future policies. An example from KV is the Flooded Meadows Project that resulted from a debate among civil servants and conservation groups about declining ecosystem services from the wetlands and the social memory and knowledge actualized by this debate.

Social learning processes are linked to the ability of management to respond to environmental feedback and direct the coupled social-ecological system into sustainable trajectories (Berkes, Colding, and Folke 2003). Through social learning (Lee 1993) resource users develop a collective memory of experiences with resource and ecosystem management. This memory provides context for social responses to ecosystem change, increases the likelihood of flexible and adaptive responses, and seems to be particularly important during periods of crisis, renewal, and reorganization (Folke et al. 2005). It draws on experience but allows for novelty and innovation within the framework of accumulated experience, referred to as framed creativity (Folke, Colding, and Berkes 2003).

If such memory represents an asset, it needs to be mobilized by developing links between key persons and providing a direction for adaptive governance. This has been referred to as building social capacity for resilience in social-ecological systems (Folke, Colding, and Berkes 2003), and it requires evoking change in social structures. For instance, in a study of the U.S. Fish and Wildlife Service Danter et al. (2000) highlight the need for organizational change as a component of ecosystem management and puts forward the role of leadership in actively initiating change within organizations. Blann, Light, and Musumeci (2003) stress the key role of social networks of practitioners in which support, trust, and sharing of lessons learned can facilitate processes of change at multiple scales.

Indeed, trust building appears to be crucial for collaboration and accessing social memory. Trust lubricates collaboration (Pretty and Ward 2001) and is a fundamental characteristic in social self-organizing processes for ecosystem management (e.g., Brown et al. 2001). A lack of trust between people is a barrier to the emergence of collaborative arrangements (Baland and Platteau 1996). Trust building was also the central component in the emergence of KV and is still crucial for the EKV approach. For instance, the Crane Project suggests how key farmers, some with experience from the flooded meadows project, were quickly mobilized into a trust-building process with conservationists and bird-watchers.

The emergence of Kristianstads Vattenrike illustrates a mental shift, a transformation toward adaptive comanagement (Olsson, Folke, and Hahn 2004). It represents a new trajectory in which ecosystems are seen as dynamic and rich, producing an array of ecosystem services. Our research on

KV suggests that much can be done within existing institutional arrangements; the technical and institutional resources for adaptive governance of social-ecological systems are already in place in a country like Sweden. Present economic incentives and other institutional arrangements both impede and enable changes toward ecosystem management. What is special about KV, and what seems to be lacking at many other places, is the social capacity to respond to ecosystem change in a way that seems to sustain and possibly enhance the capacity of ecosystems to generate essential ecosystem services. "Social capacity" in this context includes organizational capacity (stewards and organizations capable of adapting management to ecosystem dynamics) and institutional capacity to develop rules and norms that facilitate adaptive governance of social-ecological systems. This is all related to the capacity to develop social networks for horizontal and vertical collaboration. The "adaptive" aspect involves mobilizing the social memory of experiences with resource and ecosystem management when designing experiments and interpreting feedback (see figure 4.1).

Social capacity can also be understood in terms of capital, especially social capital (trust, skills in collaboration, and conflict resolution; Pretty 2003), human capital (advancement in different knowledge systems), and cultural capital (including beliefs about how people, nature, and society are related, sometimes called a worldview; Berkes and Folke 1992; Adams et al. 2003). Enhancing social capital for adaptive governance of social-ecological systems is not as simple as merely supporting local stewards and entrepreneurs. In addition, there seems to be a need for a bridging organization or a key person who provides similar functions and connects people, networks, and organizations across levels. The achievements in Kristianstads Vattenrike are not related to changes in legislation and, hence, the development of the new governance system illustrates that a lot can be accomplished when ecological knowledge is enhanced, social norms and worldviews are tuned toward an appreciation of ecosystem services, and when the realization emerges that investments in ecosystems can have positive impacts on other societal sectors as well.

The social memory of past changes in ecosystems as well as their responses can be mobilized through social networks, locally and across scales; experiential local knowledge can be fed into processes where governance of ecosystem management is decided, management practices worked out, and conflicts resolved. This requires leadership by key persons at various organizational levels. Bridging organizations and the learning they facilitate thrive under open institutions (Shannon and Antypas 1997), which provide flexibility and space for dealing with the ambiguity of multiple objectives. Another good aspect of a policy community is that people do not need to represent their organizations when they engage in informal collaboration (Westley 1995). Farmers in KV feel they can discuss management issues on equal terms with the staff from the County Administrative Board, and this has enhanced opportunities to identify common interests (Hahn et al. 2006).

In this chapter we have synthesized our empirical findings from three years studying Kristianstads Vattenrike, focusing on the role of social networks as sources of resilience in complex adaptive social-ecological systems. We conclude that social networks represent a new type of governance that provides bridges between people from different organizations or stakeholder groups. Networks can be very useful for knowledge generation, mobilizing people, identifying common interests, and starting projects. However, networks are no panacea for enhancing resilience; it all depends on the content of the exchange and what is communicated within the networks and the direction of action. One lesson learnt from Kristianstads Vattenrike is that a clear vision and firm direction of ecosystem-based development, as defined in 1989, can coexist with an exceptionally flexible project organization.

Another lesson concerns the mental shift, illustrated by the change in perceptional maps from viewing the area as water sick to water rich. EKV has emphasized communication of the ecological and cultural-historical values related to the flooded meadows; this required knowledge gained from monitoring as well as inventories, but without active communication through networks and media there would be little public attention and value formation, which is often needed for conflict resolution at a deeper level.

Networks for local collaboration require a lot of personal skills, with trust building central. Leadership at different organizational levels is required, but rather than waiting for leaders with suitable interests to turn up, EKV has searched for potential key persons that represent different stakeholder groups and organizations and built trust with them. In this way EKV has developed a policy community involving a lot of vertical collaboration. This requires skills in navigating legal, political, and financial systems as well as incorporating knowledge from external sources such as scientific knowledge and international experiences.

Social memory seems to play an important role in the self-organization process when key persons draw on social memories of several scales in the reorganization following change. Social networks can be key mechanisms for drawing on social memory at critical times, enhancing information flow across scales. The interplay between informal collaboration and formal institutions is an area that deserves more attention. Lessons from Kristian-stads Vattenrike suggest that learning and mobilization in projects for eco-system management thrive in informal settings like social networks where participation is voluntary. The challenge is to formalize these achievements, to make them less vulnerable to internal or external change or, in other words, to build resilience of desirable and sustainable trajectories.

NOTES

1. Pollinators, grazers, decomposers, and other functional groups of species in a system contribute to the performance of this system. If a whole functional group is lost locally, or if one species performing a new function enters, the functions of the system may shift dramatically (Folke et al. 2004).
2. The service metaphor was borrowed from economics to make the value of nature more visible to decision makers (Daily 1997).
3. http://www.kristianstad.se/aktuellt/tk/arkiv/402/naturvard.asp.
4. http://www.biodiv.org/decisions/default.aspx?m = COP-07 = 7748≤ = 0.
5. http://www.unesco.org/mab/about.htm.

REFERENCES

Adams, W. M., D. Brockington, J. Dyson, and B. Vira. 2003. "Managing Tragedies: Understanding Conflict Over Common Pool Resources." *Science* 302:
1915–1916.

Adger, W. N. 2003. "Social Capital, Collective Action and Adaptation to Climate
Change." *Economic Geography* 79:387–404.

Adger, W. N., and E. Tompkins. 2004. "Does Adaptive Management of Natural Resources Enhance Resilience to Climate Change?" *Ecology and Society* 9(2): 10.

Baland, J. M. and J. P. Platteau. 1996. *Halting Degradation of Natural Resources: Is
There a Role for Rural Communities?* Oxford: Oxford University Press.

Berkes, F., C. Folke, and J. Colding, eds. 2003. *Navigating Social-Ecological Systems: Building Resilience for Complexity and Change.* Cambridge: Cambridge
University Press.

Berkes, F., and C. Folke. 1992. "A Systems Perspective on the Interrelations Between
Natural, Human-Made and Cultural Capital." *Ecological Economics* 5:1–8.

Blann, K., S. Light, and J. A. Musumeci . 2003. "Facing the Adaptive Challenge:
Practitioners' Insights from Negotiating Resource Crisis in Minnesota." In
F. Berkes, and C. Folke, J. Colding, eds., *Navigating Social-Ecological Systems:
Building Resilience for Complexity and Change*, pp. 210–240. Cambridge: Cam-
bridge University Press.

Boyle, M., J. Kay, and B. Pond. 2001. "Monitoring in Support of Policy: An Adaptive Ecosystem Approach." In T. Munn, ed., *Encyclopedia of Global Environmental Change*, 4:116–137. New York: Wiley.

Brown, K., W. N. Adger, and E. Tompkins, P. Bacon, D. Shim, and K. Young. 2001.
"Trade-off Analysis for Marine Protected Area Management." *Ecological Economics* 37(3): 417–434.

Cash, D. W. and S. C. Moser. 2000. "Linking Global and Local Scales: Designing Dynamic Assessment and Management Processes." *Global Environmental
Change* 10:109–120.

Danter, K. J., D. L. Griest, and G. W. Mullins, and E. Norland. 2000. "Organizational Change as a Component of Ecosystem Management." *Society and Natural Resource* 13:537–547.

Dietz, T., E. Ostrom, and P. C. Stern. 2003. "The Struggle to Govern the Commons." *Science* 302:1902–1912.

Eckerberg, K., and M. Joas. 2004. "Multi-level Environmental Governance: A Concept Under Stress?" *Local Environment* 9(5): 405–412.

Folke, C., S. Carpenter, B. Walker, M. Scheffer, T. Elmqvist, L. Gunderson, and C. S. Holling. 2004. "Regime Shifts, Resilience, and Biodiversity in Ecosystem Management." *Annual Review of Ecology Evolution and Systematics* 35:557–581.

Folke, C., J. Colding, and F. Berkes. 2003. "Synthesis: Building Resilience and Adaptive Capacity in Social-Ecological Systems." In F. Berkes, C. Folke, and J. Colding, eds., *Navigating Social-Ecological Systems: Building Resilience for Complexity and Change*, pp. 352–387. Cambridge: Cambridge University Press.

Folke, C., T. Hahn, P. Olsson, and J. Norberg. 2005. "Adaptive Governance of Social-Ecological Systems." *Annual Review Environment and Resources* 30:441–473.

Gadgil, M., P. Olsson, F. Berkes, and C. Folke. 2003. "Exploring the Role of Local Ecological Knowledge for Ecosystem Management: Three Case Studies." In F. Berkes, C. Folke, and J. Colding, eds., *Navigating Social-Ecological Systems: Building Resilience for Complexity and Change*, pp. 189–209. Cambridge: Cambridge University Press.

Granovetter, M. 1973. "The Strength of Weak Ties." *American Journal of Sociology* 78:1360–1380.

Gunderson, L. 1999. "Resilience, Flexibility and Adaptive Management: Antidotes for Spurious Certitude?" *Conservation Ecology* 3:7.

Haas, P. M. 1992. "Epistemic Communities and International Policy Coordination." *International Organization* 46(1): 1–35.

Hahn, T., P. Olsson, C. Folke, and K. Johansson. 2006. "Trust-Building, Knowledge Generation, and Organizational Innovations: The Role of a Bridging Organization for Adaptive Co-management of a Wetland Landscape Around Kristianstad, Sweden." *Human Ecology* 34(4): 573–592.

Imperial, M. T. 2005. "Using Collaboration as a Governance Strategy: Lessons from Six Watershed Management Programs." *Administration and Society* 30: 281–320.

Kettl, D. F. 2000. "The Transformation of Governance: Globalization, Devolution, and the Role of Government." *Public Administration Review* 60:488–497.

Kingdon, J. W. 1995. *Agendas, Alternatives, and Public Policies*. New York: Harper Collins.

Lee, K. N. 1993. *Compass and Gyroscope: Integrating Science and Politics for the Environment.* Washington, DC: Island.

Lee, M. 2003. "Conceptualizing the New Governance: A New Institution of Social Coordination." Presented at the Institutional Analysis and Development Mini-Conference, May 3 and 5, 2003, Indiana University, Bloomington, Indiana.

McCay, B. J. 2002. "Emergence of Institutions for the Commons: Contexts, Situations, and Events." In E. Ostrom, T. Dietz, N. Dolsak, P. Stern, S. Stonich, and E. U. Weber, eds., *The Drama of the Commons.* Washington, DC: National Academy Press.

McIntosh, R. J. 2000. "Social Memory in Mande." In R. J. McIntosh, J. A. Tainter and S. K. McIntosh, eds., *The Way the Wind Blows: Climate, History, and Human Action,* pp. 141–180. New York: Columbia University Press.

Mintzberg, H. 1979. *The Structuring of Organizations: A Synthesis of the Research.* Englewood Cliffs, NJ: Prentice-Hall.

Newman, L. and A. Dale. 2005. "Network Structure, Diversity, and Proactive Resilience Building: A Response to Thompkins and Adger." *Ecology and Society* 10(1):2.

Norgaard, R. 1996. "Metaphors We Might Survive By." *Ecological Economics* 15:129–131.

Olsson, P., and C. Folke. 2001. "Local Ecological Knowledge and Institutional Dynamics for Ecosystem Management: A Study of Lake Racken Watershed." *Ecosystems* 4:85–104.

Olsson, P., C. Folke, and F. Berkes. 2004. "Adaptive Comanagement for Building Resilience in Social-Ecological Systems." *Environmental Management* 34(1): 75–90.

Olsson, P., C. Folke, V. Galaz, T. Hahn, and L. Schultz. 2007. "Enhancing the Fit Through Adaptive Comanagement: Creating and Maintaining Bridging Functions for Matching Scales in the Kristianstads Vattenrike Biosphere Reserve Sweden. *Ecology and Society* 12(1): 28. http://www.ecologyandsociety.org/vol12/iss1/art28/.

Olsson, P., C. Folke, and T. Hahn. 2004. "Social-Ecological Transformation for Ecosystem Management: The Development of Adaptive Co-management of a Wetland Landscape in Southern Sweden." *Ecology and Society* 9(4): 2.

Olsson, P., H. Gunderson, S. R. Carpenter, P. Ryan, L. Lebel, C. Folke, and C. S. Holling. 2006. "Shooting the Rapids—Navigating Transitions to Adaptive Governance of Social-Ecological Systems." *Ecology and Society.*

Ostrom, E. 1998. "Scales, Polycentricity, and Incentives: Designing Complexity to Govern Complexity." In L. D. Guruswamy and J. A. McNeely, eds., *Protection of Global Biodiversity: Converging Strategies,* pp. 149–167. Durham: Duke University Press.

Ostrom, E. 2005. *Understanding Institutional Diversity*. Princeton: Princeton University Press.

Ostrom, E., J. Burger, C. B. Field, R. B. Norgaard, and D. Policansky. 1999. "Sustainability—Revisiting the Commons: Local Lessons, Global Challenges." *Science* 284:278–282.

Pinkerton, E. 1989. *Co-operative Management of Local Fisheries: New Directions for Improved Management and Community Development*. Vancouver: University of British Columbia Press.

Pretty, J. 2003. "Social Capital and the Collective Management of Resources." *Science* 302:1912–1914.

Pretty, J. and H. Ward. 2001. "Social Capital and the Environment." *World Development* 29:209–227.

Scheffer, M., F. Westley, and W. Brock. 2003. "Slow Response of Societies to New Problems: Causes and Costs." *Ecosystems* 6:493–502.

Schultz, L., C. Folke, and P. Olsson. 2007. "Enhancing Ecosystem Management Through Social-Ecological Inventories: Lessons from Kristianstads Vattenrike, Sweden." *Environmental Conservation* 34 (2): 140–152.

Schultz, L., P. Olsson, Å. Johannessen, and C. Folke. 2004. "Ecosystem Management by Local Steward Associations: A Case Study from 'Kristianstads Vattenrike,'" the Swedish MA. http://www.millenniumassessment.org/documents/bridging/papers/schultz.lisen.pdf. MA Conference Bridging Scales and Epistemologies, March 17–20, Alexandria.

Shannon, M. A. 1998. "Social Organizations and Institutions." In R. J. Naiman and R. E. Bilby, eds., *River Ecology and Management: Lessons from the Pacific Coastal Ecoregion*, pp. 529–552. New York: Springer.

Shannon, M. A., and A. R. Antypas. 1997. "Open Institutions: Uncertainty and Ambiguity in Twenty-First-Century Forestry." In K. A. Kohm and J. F. Franklin, eds., *Creating a Forestry for the Twenty-first Century: The Science of Ecosystem Management*, pp. 437–445. Washington, DC: Island.

Stoker, G. 1998. "Governance as Theory: Five Propositions." *International Social Science Journal* 50(155): 17–28.

Westley, F. 1995. "Governing Design: The Management of Social Systems and Ecosystems Management." In L. H. Gunderson and C. S. Holling, eds., *Barriers and Bridges to the Renewal of Ecosystems and Institutions*. New York: Columbia University Press.

Westley, F., S. Carpenter, W. A. Brock, C. S. Holling, and L. H. Gunderson. 2002. "Why Systems of People and Nature Are Not Just Social and Ecological Systems." In L. H. Gunderson and C. S. Holling, eds., *Panarchy: Understanding Transformations in Human and Natural Systems*, pp. 103–119. Washington, DC: Island.

INTRODUCTION TO PART 3
INFORMATION PROCESSING

THE FIRST SECTION of this book is about the roles of variability in space, variability through time, and component diversity in complex systems. The second section takes a slightly different perspective and considers the same set of problems in terms of connectivity, visualizing the topology of complex systems as networks of interacting components. In this third section we again shift perspective to consider how system components, or nodes in the network, respond to signals that are received from the environment via links from other nodes.

Information processing is a general term for the responses of complex systems to external inputs. It is widely used in physics, cognitive psychology, philosophy, and information theory. In biology, and especially in ecology, ideas about information processing are captured by the terminology of functional responses (Holling 1959) and behavioral ecology. We regard information processing as a general concept that promises to unify a set of parallel ideas across several different disciplines.

In this volume we are primarily concerned with three aspects of information processing. First, the social or ecological components of complex systems often exhibit nonlinear responses, which can induce self-reinforcing feedbacks that may in turn create the potential for alternative stable states. The existence of multiple attractors for a complex system has large implications for our ability to predict change and act on our predictions. Second, we focus on human decision making, particularly in relation to the difficulties of making decisions in a world that is considerably more complex than any mental representation of it. Third, we are interested in questions of novelty and discovery and, in particular, in understanding how the processing of familiar information can lead to novel responses and the discovery of new information.

In engineering, information processing involves an input signal entering a system and a resulting output signal or response. If responses are linear, then feeding in more of the input signal gives proportionally more of the output signal. This simple relation does not hold in nonlinear systems, which often exhibit more complex dynamics and may respond to inputs in surprising ways, making them difficult to manage effectively. While chapters 5 and 6 have a strong bias toward human social systems, many of these arguments also apply to chemical and physical systems. For instance, phosphate recirculation by freshwater sediments in shallow lakes exhibits a nonlinear response to oxygen levels. If the water above the sediment layer is oxygenized, freshwater systems may bind phosphate loadings (input signal) and deposit them in the sediments so that little phosphorus is retained in the open water (output signal). Increasing the amount of phosphate in the water column load does not solicit a linear response in phosphate availability in the water column. When the oxygen above the sediment has been sufficiently depleted by the decomposition of algae, sediments will no longer retain phosphate and instead start to pump phosphate out into the water column, leading to eutrophication. Lowering phosphate loadings to the lake may have little effect, since the majority of phosphate is now provided by the sediments instead of by runoff from land. This is an example of a regime shift, and one of the conditions that needs to be fulfilled for it to occur is that some critical components in the system have nonlinear responses to another critical variable, such as oxygen or temperature.

Today we know that the potential for regime shifts is common in social-ecological systems. Linear systems are relatively easy to manage if actions with negative consequences can be identified and reduced. A more difficult problem arises when systems have the potential for regime shifts, particularly if the threshold that separates alternate system states is close to the point at which the utility gained is at an optimum. The focus of society on optimal exploitation of ecosystems often pushes natural systems close to these thresholds, leading to potentially disastrous and costly collapses (e.g., Myers and Worm 2003). While most natural systems are reactive in their responses, humans have the capacity to act proactively; but this does not necessarily benefit us if we cannot predict the behavior of the system. In chapter 6 Brock, Carpenter, and Scheffer propose a fundamentally new way of capturing signals about the environment and using them for decision making. They explore a general class of nonlinear functions, representative of dynamics in SESs, and find that, in general, as systems move nearer to internal thresholds, they start to amplify environmental variability. This means that the variation in some measurable variable might be used as an indicator of an approaching threshold (and likely system state change), signaling that caution and action are needed. For some systems, such as lakes, such warning signals could signal a transformation many years in advance.

Decisions about how to manage natural resources are closely linked to system dynamics. Decision making is the cognitive process by which a course of action is selected from multiple alternatives. It depends heavily on available information, including external signals and previous experiences. A particularly interesting aspect of decision making concerns how complex systems can cope with an information-rich system (such as the real world) while in possession of limited processing (cognitive) capacity. This is the central theme of chapter 5. Which signals need to be captured, what kind of responses are appropriate, and how should we act proactively to improve our future? Although many of these issues have been discussed under the concept of bounded rationality (Gigerenzer and Selten 2001), they are applied here in new ways to the management of natural resources.

Much of contemporary decision theory in economics is based on the desires of individuals. Decisions are based on relative preferences for the expected outcomes of different actions, but current preferences are not always a good guide to people's longer-term goals and may change substantially through time. Should one really make decisions based on today's preferences, or is it possible to account for the fact that preferences may change in the future as the outcomes of today's actions become evident? In some sense the anticipation of future needs and future changes in preferences lies at the heart of proactive behavior. From this perspective much of today's decision making in natural resource management seems very reactive. Regional and global agendas to develop sustainable use of natural resources have yet to truly grapple with the question of whose preferences should drive decision making and how these preferences may change into the future. One useful step in looking toward future preferences is scenario planning, which can help stakeholders broaden the set of possible outcomes they consider (e.g. chapters 7, 8).

Changes in preferences are closely related to the third major topic of the information processing section, that of novelty and discovery. Humans have the capacity to make planned innovations while grappling with decisions.

By this we mean that they can consciously find novel solutions to a problem. The generation of novelty involves finding new (i.e., previously not considered) signals, recombining existing information in new ways, and/or discovering new actions that achieve a desired effect. One way to foster novelty and to incorporate it into the decision-making process is to bring together people that have diverse sets of processing capabilities and different experiences, as discussed in chapters 4 and 8. Anderies and Norberg (chapter 5) propose that by diversifying the range of perspectives under consideration in solving a problem, the actors in group exercises are effectively increasing the dimensionality of their mental models of the real world—much as we hope that this book will do for its readers. Although much remains to be understood about information processing in social-ecological systems, the chapters in this section provide an interesting set of perspectives that we hope will stimulate further research into this important and exciting topic.

REFERENCES

Brock, W. 2004. "Tipping Points, Abrupt Opinion Changes, and Punctuated Policy Change." http://www.ssc.wisc.edu/~wbrock.

Gigerenzer, G., and R. Selten. 2001. *Bounded Rationality: The Adaptive Toolbox.* Cambridge: MIT Press.

Holling, C. S. 1959. "Some Characteristics of Simple Types of Predation and Parasitism." *Can. Entomol* 91:385–398.

Myers, R. A., and B. Worm. 2003. "Rapid Worldwide Depletion of Predatory Fish Communities." *Nature* 425:280–283.

5

THEORETICAL CHALLENGES
INFORMATION PROCESSING AND NAVIGATION
IN SOCIAL-ECOLOGICAL SYSTEMS

John M. Anderies and Jon Norberg

IN THIS CHAPTER we will address several theoretical challenges associated with the application of complex systems ideas to policy design and management in social-ecological systems. A complex adaptive systems (CAS) perspective on social-ecological systems involves the following key principles: they are characterized by constant change—there is no long-run optimal configuration toward which the system must be driven; diversity of components within the social-ecological system (SES) is important for its function; important interactions that structure SESs occur across multiple scales, and particularly important processes are those that drive self-organization (Levin 2003).

According to Folke and colleagues (Folke, Colding, and Berkes 2003; Folke et al. 2005), a CAS perspective requires societies to 1. learn to live with change and uncertainty, 2. combine different types of knowledge throughout the learning process, 3. create opportunities for self-organization toward social-ecological resilience, and 4. nurture the development of capacity for renewal and reorganization. We will focus here on some fundamental theoretical challenges associated with operationalizing the CAS approach with regard to how individuals and social groups processing information about their system are used to making decisions.

The concept of regimes can be combined with the concept of resilience to study change in complex SESs. Questions associated with *sustainability* can be posed in terms of the ability of an SES to remain in a desirable *regime* (or basin of attraction). The ability to remain within a regime is referred to as the resilience of the system. An important aspect of this ability is the capacity to adapt either before, during, or after a regime shift. Adaptive capacity is determined by available options in combination with the ability to use/implement these options. Obviously, the ability to generate novelty is an important aspect of adaptive capacity as it provides a source of new options, and this brings us back to the appeal of the complex adaptive systems metaphor as applied to SESs. If we are to adopt the strategy of identifying different regimes of behavior in complex SESs and asking questions about how such systems either remain in (resilience) or develop the capacity to move within (adaptive capacity) or between these regimes in the face of change (transformative capacity) we can identify several main topics that must be addressed. The remainder of the chapter is structured around these topics in the context of CAS.

Embracing complexity means continuously reminding ourselves that our understanding of SESs is based on simple representations of the "real" system. This continuous reminder helps avoid falling into the trap of beginning to believe that simple models accurately represent the system, thus using them as blueprints for decision making. Learning which factors determine the path of development of an SES requires combining diverse types of knowledge. Society must foster a process of exploration, experimentation, and evaluation (Walters 1986; Folke, Colding, and Berkes 2003).

Identifying change concerns the problem of system identity. When does a system change from one into another and what are the implications for when and how society should act? If CAS are characterized by continuous change, when is change *real* change rather than just a change in perspective? The problem of system identity has implications for avoiding shifting baseline syndromes (Pauly 1995) and preparing systems to cope with crisis.

Understanding the system involves learning how actions and endogenous feedbacks between system variables affect the future dynamics of the system. Mental representations generated by observation, reflection, and experimentation performed by several actor groups (e.g., scientists, resource appropriators, and other stakeholders) are used to map actions to expected outcomes. By what means is a configuration of a social-ecological system located in the space of all possible configurations (state space). That is, how do we know where we are?

Identifying options asks how we can systematically study the role of novelty in social-ecological systems. Novelty can be thought of as an a posteriori recognition that something has changed in the system. If so, how can novelty be incorporated into a "steering" strategy? Novelty might mean exploring multiple dimensions (rather than only aspects of one dimension) of the system and expanding the set of possible actions (Walker et al. 2002).

This typically results in alternatives to management based on the control of a single variable such as catch quotas (Acheson and Wilson 1996).

Making decisions is an important selective process in social systems. Decisions involve one or more people that take part in or have social influence on the decision process as well as the institutional framework by which decisions are governed. Decisions include evaluation of the expected outcome of each of a set of options and the "value" of these outcomes. However, prior to this step an evaluation of the cost of assessing options and outcomes determines the decision heuristics that will be used. Feedbacks between outcomes of decisions, preferences that lead to these decisions, and the decision process itself can lead to attractors that define social regimes.

EMBRACING COMPLEXITY

A worldview based on complex adaptive systems theory has several implications for how one observes, understands, and acts within a social-ecological system. Three resultant key issues when embracing a complex view of the world that we will elaborate in this chapter are 1. acknowledging that humans can only comprehend a subset of the real world (Hertwig and Todd 2004), i.e., a lower-dimensional representation of the social-ecological system that we want to navigate, and 2. realizing that self-organizing processes determine the path of development of complex adaptive systems and that much understanding can be gained by focusing on these processes rather than on the current states of the system (Levin 2003), which means that 3. on the one hand, parts of the system can change while retaining the main structures and processes but on the other, small changes can cause self reinforcing feedbacks may result in large structural changes in the system (Folke et al. 2004). We will discuss these issues further below.

The limits on human cognitive capacity (Hertwig and Todd 2004) restrict our ability to understand complex interactions. This is reflected in how humans observe the environment we live in. Over the past half century the study of the interaction between individuals, society, and natural resources has steadily advanced. A number of different threads of intellectual inquiry are important, in this respect, coming from several diverse fields, for example (in no particular order and followed by a small sample of work from each field): anthropology (Gadgil, Berkes, and Folke 1993; Lansing 2003), archaeology (Redman 2005), economics (Dasgupta and Heal

1974; Dasgupta 1982; Clark 1990; Grossman and Krueger 1995), ecology (Holling 1973), human ecology (Becker and Ostrom 1995), political science (Ostrom 1990), and geography (Adger 2000). In each of these threads the human-environment problematic is, of course, viewed distinctly, with a varying degree of emphasis on human behavior, social relations, biophysical processes, ecological processes, or economic processes. Typically, this emphasis is strongly biased toward topics of most interest in each field. For example, environmental and natural resource economists typically take institutional arrangements and infrastructure as given (no transaction costs), focus on specific policy questions (the relative merits of tradable property rights, taxes, or regulation) in a particular institutional context (highly developed market systems), and typically treat ecological processes in simplified terms. Ecologists, on the other hand, tend to focus on "pristine" ecological processes and treat human activity as a pesky disturbance. Political scientists acknowledge the central importance of institutions, but can oversimplify ecology and economics as well. This problem also translates into the policy-making arena in which simple blueprint solutions may be adopted irrespective of local or changing conditions (Wilson et al. 1994).

Not only is there the problem of diverse fields with different views, there is the fundamental problem of what framework to apply to the study of coupled social-ecological systems. At one extreme, ecologies can be seen as resources to be controlled by society (Dasgupta 1982; Clark 1990). The well-developed mathematical machinery of optimal control theory can then be brought to bear. However, there are many implicit assumptions that reduce the utility of such an approach. Perhaps the most important of these assumptions is that ecological systems are controllable—i.e., humans understand and can influence their dynamics. This is typically not the case.

It is therefore critical to move beyond using very simple "controllable systems" in the design of policy (Ostrom 2005). Ecological and social systems are far too complex to be perfectly understood, predictable, and controllable. On the other hand, we cannot admit all the complexity inherent in these systems into any reasonable analytical framework. The challenges faced in finding a middle ground includes finding methods to generate "coarse-grained" representations of complex systems. However, since SESs are CAS and undergo constant change, the crucial point is that this representation needs to be subject to change based on experimentation and continued diversication of knowledge types, e.g., by adaptive management (Holling 1978; chapter 7, this volume) and comanagement (Folke et al. 2005).

The common approach to dealing with complex systems is to treat most factors as exogenous (and typically hold them constant) and develop simple models based on a few key variables (i.e., a low-dimensional representation). Through the analysis of the resulting highly simplified model, considerable progress can be made in understanding general principles concerning the sustainability of social-ecological systems across a range of scales (e.g., Stiglitz 1974; Solow 1974; Dasgupta and Heal 1974; Hartwick 1977; Beltratti 1997; Carpenter, Brock, and Hanson 1999; Carpenter, Ludwig, and Brock 1999; Anderies 2003, 2004). Why then introduce the CAS perspective when traditional models produce such important insights? Because policies based on such general themes typically fail because they miss important details of local conditions. Put another way, context is important. The importance of context suggests that rather than relying on principles derived from optimization over a set of trajectories generated by one or two dimensional representations of systems (i.e., models with one or two state variables), policy should be based on an understanding of the selective processes that determine possible development paths of SESs.

As a characteristic of CAS, selective processes are the result of local interactions between elements (Levin 1998) and act strongly on redundant properties between system components (CH 2). This promotes the most "efficient" elements and shapes the composition of both the social and ecological systems. Elements in the context of SESs include species in the ecological domain and different activities, institutions and people in the social domain. Important selective processes occurring in SESs include competition among species (natural selection) and competition among different forms of social and institutional arrangements and economic organization (competition in markets). Attempts to steer a system must focus on which factors determine the current and potential attractors of the system (i.e., the long-run configuration of the system resulting from the selective processes acting on its components) rather than the current state or a particular trajectory. Thus, focusing on self-organization involves understanding the sources of diversity in both ecological and social components as well as the processes that select among this diversity. These are particularly important for the resilience of any set of processes maintained by the interactions between the elements in the system because the capacity to adapt to a shift among elements without reducing overall performance of the group is largely determined by the range of responses present in that group (Elmqvist et al. 2003).

COPING WITH CHANGE

Coping with complexity and unpredictable change in the environment is nothing new, of course. In particular, economists have developed an extensive literature in this area over the past fifty years. One thread of this literature began with work on uncertainty in the demand for a publicly provided resource (Weisbrod 1964). The motivation for this work, as Arrow and Fisher (1974) point out, is that "any discussion of public policy in the face of uncertainty must come to grips with the problem of determining an appropriate attitude toward risk on the part of the policy maker" (Arrow and Fisher 1974:312). Namely, there may exist some option value in providing a public good. It is very natural to map this idea onto a particular public good: preservation of environmental resources. In fact, Arrow and Fisher (1974) have addressed this very issue and shown that there is a positive option value associated with preserving the environment in the case in which investment decisions (or social actions more generally) lead to irreversible damage. The idea of option value has since become a central means of dealing with uncertainty in environmental economics and is closely related to the more general area of investment under uncertainty and the concept of "real options" popularized by Dixit and Pindyck (1994).

Note, however, the specific set of assumptions that underlie this approach: 1. the objective is to control the system through a sequence of investment (resource use) decisions to maximize the value of a stream of future benefits, and 2. complexity is dealt with through the direct use of probabilistic models that assumes a known distribution, i.e., the focus is *risk* rather than systemic uncertainty. This approach is fundamentally about controlling a system with partial information—potentially quite different from a complex systems approach. The risk-based approach represents a complex system (with perhaps several hundred state variables) with a system of far fewer variables (five or typically less), with all the neglected variables lumped into a "random" component that impinges upon the chosen few. If the distribution of this random component is well-known, then such an approach is reasonable and falls within the domain of risk analysis.

The typical result from the traditional option value approach is that when there is a risk of an irreversible change in an ecological system that will reduce the value of the stream of goods and services it provides, precaution is called for. The less that is known about the ecology, the more precautionary society should be. The basic logic is simple: if preservation of a resource provides more future opportunities than would be available if it were not preserved, then the option value of preserving the resource will exceed the use value. Even if the many problems with this approach could be overcome, there remains the fundamental problem that SESs are not composed of a single controller making welfare-maximizing decisions about the use of environmental resources. Rather, SESs are composed of a collection of heterogeneous interacting agents—i.e., are complex adaptive systems—and as such are inherently unpredictable.

Another thread in the economics literature has developed around treating economies as complex adaptive systems (Anderson, Arrow, and Pines 1988; Arthur, Durlauf, and Lane 1997; Blume and Durlauf 2005). However, the focus of this literature is not about how to manage complex systems per se, but rather about how endogenous complexity is generated in economic systems. This work deemphasizes equilibrium analysis and focuses on nonlinear feedbacks between aggregate patterns and agent behavior (Arthur 1999). For example, when agents are heterogeneous and they change their beliefs about market predictors, very complex dynamics can emerge even in very simple models (Brock and Hommes 1997, 1998; Goeree and Hommes 2000). Some take the CAS perspective one step further to explain the behavior of a real CAS, the stock market, using agent-based financial markets (LeBaron, 2000, 2001a, b). This work provides us a window on the operation of CAS from the outside and may help uncover general principles about how they operate. However, this work is not focused on developing principles about how to manage a CAS and identify change *from within the system*. This is a quite different problem.

From an analytical perspective, embracing complexity is nontrivial. We advocate avoiding the temptation of trying to understand the complexity (or actual operation) of SESs and instead building up some understanding of the different regimes or attractors they may occupy. Regimes are important because they exploit the fact that complex systems tend not to frequently visit all possible states, but rather tend to frequently visit only a small subset of possible states. Further, they tend to move relatively rapidly between them when conditions change or perturbations push them out of one regime. We next develop some ideas about how the concept of regimes can improve system understanding and increase the degree of predictable change.

UNDERSTANDING THE SYSTEM

The fundamental sustainability problem faced by society is the fact that the system in which they are embedded has an extremely large number of variables (a formal mathematical representation would thus have many state variables or dimensions). These state variables interact to generate the landscape which society must navigate. Given the cost associated with generating an understanding of the global properties of this high-dimensional landscape, society must rely on lower-dimensional representations to navigate. A simple example of this situation is the use of a map for navigation. A map is a two-dimensional representation of a three-dimensional landscape. If a navigator wishes to move from point A to point B, she may use the map along with some *local information* (location of a star or orientation of a magnetic field) to choose a direction of motion to accomplish this task. If the map is not a topographic map, there is no way to determine whether the chosen path is, in fact, feasible. There may be a one-thousand-meter-deep chasm between points A and B along the chosen path. Knowledge of such a chasm requires information about a third dimension—namely, elevation. Of course, a topographic map has this third dimension *projected* onto the two-dimensional representation, making navigation easy.

Typically it is impossible to project all dimensions of a social-ecological system onto low-dimensional representations thereof. Necessarily, information is lost in such low-dimensional representations, which unfortunately are all that is typically available to choose a course of action (i.e., decide in which direction to move). The analogy is a map showing only the relative north-south and east-west locations (x, y coordinates) of particular objects on the landscape with no information about the elevation (z coordinate). Although such a map can be an aid in navigating from A to B, its use could lead to a much more arduous journey than one with elevation information. How does one navigate with such a map? Using local elevation information, of course. If on a certain trajectory a chasm is encountered, i.e., the change in local elevation per unit of movement becomes too large, the navigator can reorient her path to one with an acceptable elevation gradient. This is exactly the approach typically followed in social-ecological systems: pursue an action (path through state space) until gradients associated with it become too large, then change the path. This is perfectly acceptable if all necessary paths are available. Unfortunately, particular path choices can shrink the set of available path choices over time. It is this fact that moti-

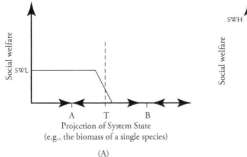
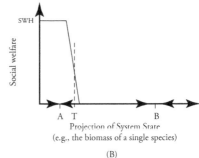

FIGURE 5.1. A depiction of the welfare associated with different stable attractors. (A) lower intensity of use, which allows for a larger equilibrium biomass. (B) Higher intensity of use resulting in a lower equilibrium biomass. As intensity of use increases, welfare may increase, but the size of the attractor (regime) associated with that level of welfare may shrink.

vated Arrow and Fisher's early work on the option value of environmental preservation (Arrow and Fisher 1974), which attempts to determine the value of keeping paths open.

To continue with the map analogy, policy actions based on CAS approaches and resilience take as their primary objective to not fall off an edge rather than to get from point A to point B most efficiently. This means enlarging the size of desirable basins of attraction. However, even this strategy faces the same problems when a low-dimensional representation of a higher-dimensional system is used for this purpose. It is often the case that intensified use of a particular ecosystem service, while generally enhancing welfare in the short to medium term, can cause the size of the basin of attraction associated with the service in question to shrink (figure 5.1).

Figure 5.1 (A) shows the relationship between social welfare and a one-dimensional projection of the system state. This one-dimensional projection might be the biomass of a species of particular interest to humans. The actual system consists of numerous interacting species (many dimensions), but the biomass of only one species (one-dimension) is used to represent the state of the system. Points A and B are stable equilibrium states and point T represents the location of the boundary between the basins of attraction of each of the stable equilibria. As long as the system remains in the basin of attraction of steady state A, society enjoys a social welfare level of SWL (social welfare of low intensity use). Further the system can tolerate a perturbation to the state variable on the x-axis from equilibrium of size $|T\text{-}A|$ and still provide welfare level SWL. If the system is perturbed further, it will move into the domain of attraction of the point B. While in this domain of attraction, the system produces no social welfare. Now consider intensifying the use of some service that flows from the ecological system to increase social welfare (figure 5.1 [B]). An unintended consequence of such intensification is that the domain of attraction of point A shrinks (Anderies, Janssen, and Walker 2002; Carpenter, Ludwig, and Brock 1999). So although the system is producing a higher level of benefits for society (SWH), the system cannot withstand as large a perturbation as in the low-intensity use case before moving into the domain of attraction of point B where the system produces no benefits to society.

Now suppose a society is in the situation depicted in figure 5.1 (B) and has chosen as a management strategy to enhance the resilience of that configuration to perturbations of the variable on the x-axis. There are two challenges here: actually measuring the size of the basin of attraction of point A (Carpenter et al. 2001) and determining how actions aimed at increasing the size of the basin of attraction affect the system variables that have been neglected. Supposing society has dealt with the first challenge, and had discovered a mechanism to expand the size of the basin of attraction around point A, how might the neglected variables respond? One possibility is depicted in figure 5.2.

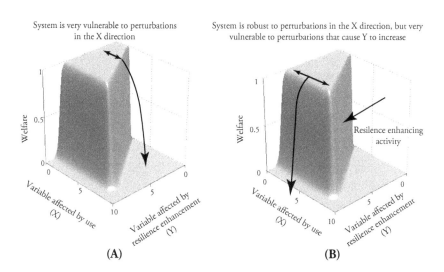

FIGURE 5.2. A depiction of the way resilience-enhancing activities may affect the vulnerability of the system to fluctuations in neglected variables. Resilience may be enhanced to perturbation in the X directions, but vulnerabilities may be introduced to perturbations in the Y direction.

When the variable affected by resilience-enhancing activities, Y, is zero, the situation shown in figure 5.1 (B) obtains (figure 5.2 [A]). When Y is low, a small perturbation in the X direction can cause the welfare generated by the system to fall as shown by the arrows. If society undertakes resilience-enhancing activities (increases Y), the size of the basin of attraction of the desirable state can be enhanced (figure 5.2 [A]). However, increasing Y may move the system toward a different threshold (figure 5.2 [B]). Now a small perturbation in the Y direction can cause the system to move into a different undesirable state that generates no welfare. This trade-off of one vulnerability for another is well known in engineering (Csete and Doyle 2002; Boyd and Barratt 1991; Gilbert and Kolmanovsky 1999). In some sense there is a conservation law that suggests that fragility can only be moved around in a given system, not eliminated completely. This idea has been used to develop a theory of design limits for optimal policy (Brock and Durlauf 2004).

If the dynamics of the system are well understood, this conservation of fragility property is not necessarily a problem. Society can assess the relative sizes of expected perturbations in an attractor of particular dimensions and make a balanced decision about where to place the system given some characterization of the social attitude toward risk. If, for example, the expected perturbations in the Y direction are small and those in the X direction are large, then the position depicted in figure 5.2 (A) may be a reasonable choice. The problem emerges when the entire Y dimension is unaccounted for! That is, what may seem a perfectly sound and conservative strategy of enhancing the resilience of the system using a model based on a lower-dimensional projection of the system may actually reduce the resilience of the system when the additional dimension is considered.

There are three avenues available to address this problem. The first is well known: exploit local information to infer global behaviors. This is an extremely difficult task in general but is better than nothing. The key is to develop measurement and experimental protocols to probe for thresholds in social ecological systems. One such protocol is the adaptive management approach developed by Walters (1986). Adaptive management advocates probing for boundaries through controlled "policy as experiment" activities. Unfortunately, conducting experiments at the right scale is problematic; it may involve the loss of livelihoods for extended periods. Unless society can compensate for such periods, adaptive probing is unlikely to occur except in models of actual systems (Walters 1997).

The second avenue involves reflecting on the decision sequence that brings a system to a certain point as it impinges on subsequent decisions. The nature of the decision to intensify resource use and move from the situation in figure 5.1 (A) to figure 5.1 (B) may condition the next decision. If the decision to intensify resource use was considered by society to be morally sound, ethical, and equitable, then the decision to attempt to enhance the resilience of this configuration by increasing Y (figure 5.2 [B]), even with the possibility of introducing new vulnerabilities, may be reasonable. If, on the other hand, the decision to intensify resource use was not morally sound, ethical, or equitable (e.g., excessive exploitation of resources by the privileged few), then the decision to increase Y should be called into question. Rather, the resilience of state A should be increased not by increasing Y but rather by reducing intensity of use and moving back to the situation pictured in figure 5.1 (A).

The final avenue involves attempting to maintain the ability to adapt to changes in system states. Thus, instead of trying to better understand the dynamics of and thus better manage a particular SES configuration, society might invest in developing an ability to generate novel ways to interact with the system or maintain a diverse set of technologies that would allow benefits to be generated under completely different ecological and social conditions. For example, rather than try to manage better (which might include enhancing resilience) at state A in figure 5.1, society might invest in developing new technology capable of generating benefits in state B. If society maintained two technologies, one for A and one for B, when it experienced a perturbation that pushed it past the threshold "T," society could simply switch technologies. The problem: maintaining multiple technologies or the ability to generate novel ways to interact with the system is costly. Further, developing means to generate novelty, without the luxury of being able to peer into the future, is difficult. Putting the concepts of novelty and diversity into practice is the second core challenge with CAS thinking in management contexts.

IDENTIFYING OPTIONS: NOVELTY AND INNOVATION

In this section we will explore the notion of innovation in the context of finding new ways to deal with a situation or discovering new variables or processes that help in understanding and affecting the dynamics of SESs.

This is a familiar aspect of everybody's experience in real life. To appreciate the problem of understanding what innovation means, put yourself into the position of "creator." For those readers who have a problem with this, you may envision that you have created a computer program that mimics some system by defining which variables interact in what ways. This means you, as the creator, have defined what is observable and the interactions between what is observable, leaving little room for innovation and surprise. Thus, when it comes to learning about the workings of a dynamic complex system, innovation lies in the possibility to expand the limited perspective of the agents or, in mathematical terms, increase the dimensionality of the agents' mental representation in relation to the "real system." We will focus here on the issue that the dimension of observable states, available actions, and considered outcomes is far lower than the dimension of the real system and discuss how novelty and innovation can be thought of in terms of expanding this limited "dimensionality" by bringing different mental models together.

CHALLENGES TO THE STUDY OF INNOVATION

The terms *novelty* and *innovation* appear frequently in economics, evolutionary biology, and ecology with respect to social, biological, and ecological systems. Unfortunately, the meanings associated with these terms are often vague and the strict definition of novelty found in many dictionaries is not particularly useful. Novelty is often defined in terms of innovation or something new or strange. One useful definition comes from the *Merriam-Webster's Dictionary of Law:* "quality of being different from anything in prior existence." This definition is useful because it is practical. But notice that this definition is only useful a posteriori. That is, it is only possible to determine if something is novel by comparing to a state of the system at a previous time. This is not particularly useful in trying to *plan* responses to ecological change. Another source, the *Oxford English Dictionary*, includes in its list of definitions: "hitherto unknown." This definition focuses on the state of knowledge of the observer, rather than on a state of the system, and is suggestive of how novelty may be studied.

If we wish to understand novelty using formal models, we must somehow develop a means to generate novelty using information at the present and past states. In the law dictionary definition we must go to a future state that is not known to exist at the present state and compare it with the present state. A moment's reflection suggests that, in fact, what we call novel is only a *discovery* of a new state. Thus what an agent calls novel is completely dependent on that agent's perception of the system. This attribute of novelty creates problems when trying to model it.

Consider trying to generate novelty using a deterministic algorithm (the only way to do it on a computer). One approach is to imagine that new things are constructed from building blocks. In most systems there are constraints on the possibilities for combining these fundamental elements so it is reasonable to specify a set of rules for combining elements. However, once the building blocks of assembly rules (dynamics) are defined, a generalized dynamic system has been defined. Once an initial state is specified and the system is started, there is nothing novel about the evolution of the system. The system moves through (or explores) state space. Novelty can only be experienced by an agent from within the system when the state moves into a region of state space not previously observed by the agent. In this way the agent *discovers* a region of state space. The modeler knows the full extent of the state space and thus cannot observe novelty within the model. At best, the modeler may be surprised by the dynamics of the system. Ultimately, the only way to approximate novelty is to allow for a very large state space so that it is beyond the computational ability of the modeler to fully explore the model dynamics. That is, the behavior of the model becomes unpredictable (or at least very difficult to predict).

Another source of unpredictability comes not from model complexity but from the limits to measurement. Consider the Lorenz equations, a three-dimensional dynamical system developed by atmospheric scientist Ed Lorenz in 1963 as a very simplified model of convection cells in the upper atmosphere. The dynamics of this system are extremely sensitive to initial conditions. An arbitrarily small difference in initial conditions can lead to completely different dynamics in the long run. Thus, unless the state of a system can be located "perfectly," its future behavior is unpredictable. Is its future behavior novel?

What is more, the Lorenz equations are an example of what might be a "closed" system. By definition, the trajectories are limited to three dimensions. Systems that exhibit true novelty, such as complex adaptive systems, are "open," in the sense that the dimension of the state space may increase indefinitely. Such open ended systems are difficult to place in a traditional management context because management problems are inherently closed. Typically to manage a system, the dynamics must be known, or at least the distribution of the dynamics must be known, and an explicit objective function must be stated. What is the management question if we admit we have no idea how the present set of imaginable management actions will affect the dynamics of the system in the future or what the set of management actions might look like? The obvious response is to focus attention on subcomponents of the system that are understandable and manageable and neglect the larger system. The only reasonable management question regarding the larger system is to what extent management actions regarding subcomponents of the system affect the ability of particular subcomponents to cope with unexpected change. This type of management question is often framed in terms of resilience, adaptive capacity (adaptability), and transformative capacity (i.e., transformability; Gunderson and Holling 2002; Walker et al. 2004).

Unfortunately, respecifying the problem in terms of a (better understood) closed system embedded in a larger (poorly understood) open system only partially eliminates the problem of trying to navigate an open system. First, the phrase *cope with unexpected change* implies that human well-being can be clearly defined (to measure how well it "coped"), yet it is difficult to do so for "unexpected" outcomes. Second, given that the subcomponent of the system that we wish to have the capacity to cope with change is the social system, we have the additional problem of the distribution of reduction of well-being in the face of change (e.g., in the case of climate change, most of the reduction in well-being will be borne by poorer people).

NOVELTY, INNOVATION, AND REGIME SHIFTS

To understand the relevance for SESs, consider a case where ecological processes give rise to several alternative attractors. Suppose that currently the biophysical system occupies attractor 1, and society has organized its activities around a range of biophysical goods and services that flow while in attractor 1. Suppose further that the use of these goods and services can push the state of the system toward the boundary of and/or reduce the size of attractor 1. The ability of the social system to sense when the state of the biophysical system is moving near a boundary or causing the basin to shrink and take corrective action is its adaptability. However, suppose society chooses an alternate strategy to allow for the possibility that biophysical system may leave the basin around which it is presently organized and enter a different basin of attraction. In this case society invests in the capacity to reorganize around a different set of goods and services that are available in the new basin. This capacity has been termed transformability, and has been thought of as fundamentally different from adaptability.

Whether or not these two process are different is related to the problem of system dimension (and identity, which we discussed earlier). The definition of the biophysical attractors requires focusing on a few relevant variables of the coupled SES. In practice this is done by treating some variables as parameters (holding them constant) and studying how the system behavior changes as such parameters are varied. Suppose for the sake of exposition that there is one such social variable, call it a. For each *fixed* value of a the biophysical system of dimension N has very specific dynamics. Suppose at a given time t society believes a is fixed at $a = a^*$ and organizes around the biophysical dynamics at $a = a^*$. For this fixed value of a, society may have significant capacity to alter its activities (the way it uses the stream of goods and services generated in the attractor with $a = a^*$) to keep the system in this attractor. Suppose then that society discovers that a is not fixed—that it is a variable. Now from the point of view of the observer, i.e., society, the dimension of the system has increased from N to $N + 1$. Suppose further that by varying a, society can maintain its "well-being" even if the biophysical system leaves the attractor defined at $a = a^*$ and enters another. That is, society can "transform" and still maintain its well-being in a new area of state space.

Now if society develops strategies to alter the variable a to maintain the system in either of two "subattractors" of dimension N, then the resulting feedbacks may generate a new $N+1$ dimensional attractor. So has a transformation occurred? Or has the system just discovered a new region of state space? That is, nothing transformational has occurred—only the perception of the dimension of the system has changed. The problem of dimension can thus make the metaphors of attractors, adaptation, and transformability misleading. By diverting attention to whether one type of change is adaptation or transformation, more important issues may be missed. Why? Because lower-dimensional attractors, some of which are seen as desirable and some of which are seen as undesirable, may be subsumed into a higher-dimensional attractor that is desirable.

MAKING DECISIONS

Deliberate human actions are based on reasoning, judgment, and decision making (Hertwig and Todd 2004). But, as we have discussed in the previous sections, the "real" environment is often too complex to grasp, and so a simpler representation of the environment is used to make decisions. When faced with complex decisions, humans take shortcuts, i.e., they tend to use decision heuristics (Janssen and Jager 2001). For example, as Hogarth and Karelaia (2004) point out:

"There is considerable evidence that people have difficulty facing trade-offs and often use choice strategies that avoid them" (see, e.g., Payne, Bettman, and Johnson 1992). *Two reasons are typically advanced. One is cognitive. Trade-offs are difficult to execute. The other is emotional. People don't like to face trade-offs explicitly* (Hogarth 1987; Luce, Bettman, and Payne 1997). *In addition, people often do not have well-formed preferences and construct apparent preferences in the act of choosing* (Bettman, Luce, and Payne 1998; Slovic 1995).

One can think of a decision as being the outcome of processing information in three domains: 1. the available options, 2. the potential outcomes, i.e., the effects of the alternative actions on the SES, and 3. their preferences for those outcomes. The more options and potential outcomes, the more complex the decision, and the more effort it takes to obtain the information needed.

When does it make sense to make a complex decision that may be very costly? When would it be better to reduce the cognitive cost by, for example, only considering one outcome or only a few possible actions? Payne (1976) and Beach (1990) describe how combined decision strategies typically start with the elimination of a number of alternatives, followed by a stage in which the remaining set of alternatives is scrutinized in more detail. What is the rationale leading to the choice of these remaining alternatives? How can an agent predict the expected benefit of increasing the number of actions or outcomes to consider in the decision process? There might be little or no prior knowledge (or at least very high uncertainty) about what benefits new actions will generate or whether the correct set of outcomes is being considered. Is it worth the effort finding the needed information? It has been suggested that the degree of relative current satisfaction (satisfaction relative to previous experiences or relative to other agents) plays an important role in this regard (Janssen and Jager 2001). If human well-being deteriorates over time, it may become clear that past descisions were poor and other actions need to be considered or that particular outcomes were weighted too heaviliy. Without a certain solution, increasing the base of alternatives considered or relying on advice from others in making decisions may be the only alternative. In this case bridging organizations, NGOs, or local stewards may substantially reduce the information-processing costs by providing easily accessible and preprocessed information. However, this may come at a cost of reduced diversity of information available for decision making and may reduce the capacity of agents to explore the new dimensions of the system.

When decisions are made by groups of people, as is often the case for policies regarding common pool resources, another level of complexity emerges related to the criteria used for weighting multiple participants' preferences and expected outcomes of alternative policies (Dietz, Ostrom, and Stern 2003; Ostrom 2005). Any enforced policy may generate changes in human behavior that affect multiple ecosystem services (e.g., timber harvesting affects number of trees, fruit production by trees, forest-dwelling animals, infrastructure, amenities, etc.). These services may be differently valued (different preferences) by different agents. As such, human interaction with nature challenges individuals and groups with multi-objective planning problems. Such models require explicit weighting of outcomes (Keeney and Reiffa 1976), which can be affected by social influence and power structures in the group decision process (Ibarra and Andrews 1993). Thus, even if the dynamics of the ecological system are *perfectly understood*, complexity in group decision processes may generate socially suboptimal outcomes.

A fundamental yet still undeveloped area in complexity theory is the notion of social attractors. We suggest here that a helpful means of conceptualizing social attractors may be to focus on the basis for decision making at different levels of society. For example, in southern Sweden in the Kristianstad municipality area a shift in the social attractor, conceived in terms of the basis for decision making, may have been observed in recent years. A "water sick" area (translated from the Swedish meaning worthless for production because of abundant water) just outside of the main city in the municipality was of low priority until 1989. A few people began lobbying for conservation efforts, and skilled stewardship then led to the acceptance across many organizational levels that the former water sick area is now considered a "water kingdom" (Olsson, Folke, and Hahn 2004). Today signs are emerging that the water kingdom may well become a new identity for the community. Conservation of the natural beauty of the area is now a high priority in decisions from individual landowners' land use practice to industrial relocations. Decisions at all levels (up to international when the area became a RAMSAR wetland convention area and a Man and the Biosphere reserve) are made placing a high priority on preserving and enhancing the amenity value of this area. This example also illustrates the positive feedbacks between the two configurations that are played out by processes acting on a level above individual organizations, i.e., the general view of the area and the overall beauty, which is determined by actions occurring at many levels. If one sees the social attractor as the preference that is the basis for decision making in the community, the Kristianstad example seems to suggest a marked shift in the path of development of that area, although social structure per se has remained unchanged. In this sense records of organized decision-making processes might be used to search for both slow succession and rapid shifts in bases for decision making and may thus provide historical evidence of shifts in social attractors.

In this chapter we have tried to summarize some of the fundamental challenges that arise when applying ideas from a complex adaptive systems perspective. The problems that arise are associated with how to analyze complex, open systems in a management context, identifying and studying novelty, and making decisions that influence many streams of goods and services simultaneously. We have argued that resource governance must be based on more than low-dimensional representations of social-ecological systems. However, just how to do this is elusive. We can only provide a few general heuristics:

1. Any representation of an SES will be of far lower dimension than the real system. Thus no matter what management actions are undertaken, the system may move into new regions of state space. Understanding just how this occurs is extremely difficult. However, it may be enough just to recognize that it is likely that this will occur, so that we are not lulled into believing low-dimensional representations of the system. Management decisions based on such models, whether directed at maximizing performance or enhancing resilience to particular perturbations, can erode resilience to other perturbations. Better managing the trade-offs between resilience to different classes of disturbances is an extremely difficult but important task.

2. Understanding the nature of self-organizing processes that operate in the system, such as the preferences and network of social influence that af-

fects decisions, may be as important as understanding the actual dynamics of the system. Learning how to manage based on such an understanding may be more important as human pressure on ecological systems increases.

3. As Arrow and Fisher (1974) point out, society must come to grips with determining an appropriate attitude toward risk. However, we suggest that society must go one step further and grapple with developing an appropriate attitude toward uncertainty. Complex adaptive systems thinking, although difficult to operationalize, may be of some value in helping society understand the components of such an attitude.

4. Although complex adaptive systems thinking provides a nice metaphor that is consistent with many systems, it faces many challenges in becoming a useful management tool. Namely, it emphasizes novelty, diversity of components, and selective forces which are often understood only a posteriori. Managing selective forces and fostering novelty a priori is quite a different and potentially more difficult task than understanding complex systems.

Clearly, given the analytical limitations of the complex adaptive system approach, a combination of traditional techniques based on low-dimensional representations of SESs along with high analytical power and ideas based on complex adaptive systems thinking will have to be brought to bear in future resource management decisions. Just how to bring the two together should be the focus of a vigorous research effort as the problems faced by human societies move to larger spatial scales and increasing levels of complexity.

REFERENCES

Acheson, J. M., and Wilson, J. A. 1996. "Order Out of Chaos: The Case for Parametric Fisheries Management." *American Anthropologist* 98:579–594.

Adger, N. W. 2000. "Social and Economic Resilience: Are They Related?" *Progress in Human Geography* 24:347–364.

Anderies, J. M. 2003. "Economic Development, Demographics, and Renewable Resources: A Dynamical Systems Approach." *Environment and Development Economics* 8:219–246.

Anderies, J. M. 2004. "Minimal Models and Agroecological Policy at the Regional Scale: An Application to Salinity Problems in Southeastern Australia." *Regional Environmental Change*. In Press.

Anderies, J., M. Janssen, and B. Walker. 2002. "Grazing Management, Resilience, and the Dynamics of a Fire Driven Rangeland System." *Ecosystems* 5:23–44.

Anderson, P. W., K. Arrow, and D. Pines, eds. 1988. *The Economy as an Evolving Complex System*. Redwood City, CA: Addison Wesley Longman.

Arrow, K., and A. C. Fisher. 1974. "Environmental Preservation, Uncertainty and Irreversibility." *Quarterly Journal of Economics* 88:312–319.

Arthur, W. B. 1999. "Complexity and the Economy." *Science* 284:107–109.

Arthur, W. B., S. Durlauf, and D. Lane, eds. 1997. *The Economy as an Evolving Complex System II*. Redwood City, CA: Addison Wesley Longman.

Beach, L. R. 1990. *Image Theory: Decision Making in Personal and Organizational Contexts*. Chichester: Wiley.

Becker, C. D., and E. Ostrom. 1995. "Human Ecology and Resource Sustainability: The Importance of Institutional Diversity." *Annual Review of Ecology and Systematics* 26:113–133.

Beltratti, A. 1997. "Growth with Natural and Environmental Resources." In C. Carraro and D. Siniscalcoa, eds., *New Directions in the Economic Theory of the Environment*. Cambridge: Cambridge University Press.

Bettman, J. R., M. F. Luce, and J. W. Payne. 1998. "Constructive Consumer Choice Processes." *Journal of Consumer Research* 25:187–217.

Blume, L., and S. Durlauf, eds. 2005. *The Economy as an Evolving Complex System III.* Oxford: Oxford University Press.

Boyd, S., and C. Barratt. 1991. *Linear Controller Design: Limits of Performance.* Upper Saddle River, NJ: Prentice-Hall.

Brock, W., and S. Durlauf. 2004. "Elements of a Theory of Design Limits to Optimal Policy." *Manchester School* 72:1–18.

Brock, W., and C. H. Hommes. 1997. "A Rational Route to Randomness." *Econometrica* 65:1059–1095.

Brock, W., and C. H. Hommes. 1998. "Heterogeneous beliefs and routes to Chaos in a Simple Asset Pricing Model." *Journal of Economic Dynamics and Control* 22:1235–1274.

Carpenter, S., W. Brock, and P. Hanson. 1999. "Ecological and Social Dynamics in Simple Models of Ecosystem Management." *Conservation Ecology* 3:http://life.csu.edu.au/consecol/.

Carpenter, S., D. Ludwig, and W. Brock. 1999. "Management of Eutrophication for Lakes Subject to Potentially Irreversible Change." *Ecological Applications* 9:751–771.

Carpenter, S., B. Walker, J. M. Anderies, and N. Abel. 2001. "From Metaphor to Measurement: Resilience of What to What?" *Ecosystems* 4:765–781.

Clark, C. W. 1990. *Mathematical Bioeconomics: The Optimal Management of Renewable Resources.* New York: Wiley.

Csete, M. E., and J. C. Doyle. 2002. "Reverse Engineering of Biological Complexity." *Science* 295:5560.

Dasgupta, P. 1982. *The Control of Resources.* Cambridge: Harvard University Press.

Dasgupta, P., and G. Heal. 1974. "The Optimal Depletion of Exhaustible Resources." In *Review of Economic Studies: Symposium on the Economics of Exhaustible Resources,* 41:3–28. Edinburgh: Longman.

Dietz, T., E. Ostrom, and P. C. Stern. 2003. "The Struggle to Govern the Commons." *Science* 302:1902–1912.

Dixit, A., and R. Pindyck. 1994. *Investment Under Uncertainty.* Princeton: Princeton University Press.

Elmqvist, T., C. Folke, M. Nyström, G. Peterson, J. Bengtsson, B. Walker, and J. Norberg. 2003. "Response Diversity and Ecosystem Resilience." *Frontiers in Ecology and the Environment* 1:488–494.

Folke, C., S. Carpenter, B. Walker, M. Scheffer, T. Elmqvist, L. Gunderson, and C. Holling. 2004. "Regime Shifts, Resilience, and Biodiversity in Ecosystem Management." *Annual Review of Ecology Evolution and Systematics* 35:557–581.

Folke, C., J. Colding, and F. Berkes. 2003. "Synthesis: Building Resilience and Adaptive Capacity in Social-Ecological Systems." In F. Berkes, J. Colding, and C. Folke, eds., *Navigating Social-Ecological Systems: Building Resilience for Complexity and Change,* pp. 352–387. Cambridge: Cambridge University Press.

Folke, C., T. Hahn, P. Olsson, and J. Norberg. 2005. "Adaptive Governance of Social-Ecological Systems." *Annual Review of Environment and Resources* 30:441–473.

Gadgil, M., F. Berkes, and C. Folke. 1993. "Indigenous Knowledge for Biodiversity Conservation." *Ambio* 22:151–156.

Gilbert, E.G., and I. Kolmanovsky. 1999. "Fast Reference Governors for Systems with State and Control Constraints and Disturbance Inputs." *International Journal of Robust and Nonlinear Control* 9:1117–1141.

Goeree, J.K., and C. Hommes. 2000. "Heterogeneous Beliefs and the Nonlinear Cobweb Model." *Journal of Economic Dynamics and Control* 24:761–798.

Grossman, G., and A. Krueger. 1995. "Economic Growth and the Environment." *Quarterly Journal of Economics* 110:353–377.

Gunderson, L.H., and C.S. Holling, eds. 2002. *Panarchy: Understanding Transformations in Systems of Humans and Nature.* Washington, DC: Island.

Hartwick, J.M. 1977. "Intergenerational Equity and the Investing of Rents from Exhaustible Resources." *American Economic Review* 67:972–974.

Hertwig, R., and P. M. Todd. 2004. "More Is Not Always Better: The Benefits of Cognitive Limits." In David Hardman and Laura Macchi, eds., *Thinking: Psychological Perspectives on Reasoning, Judgment and Decision Making,* pp. 213–231. New York: Wiley.

Hogarth, R. 1987. *Judgement and Choice.* 2d ed. New York: Wiley.

Hogarth, R., and N. Karelaia. 2004. *Simple Models for Multi-attribute Choice with Many Alternatives: When It Does and Does Not Pay to Face Trade-off.* Department of Economics and Business, Universitat Pompeu Fabra, Economics Working Papers.

Holling, C.S. 1973. "Resilience and Stability of Ecological Systems." *Annual Review of Ecology and Systematics* 4:1–23.

Holling, C.S. 1978. *Adaptive Environmental Assessment and Management.* London: Wiley.

Ibarra, H., and S.B. Andrews. 1993. "Power, Social Influence, and Sense Making: Effects of Network Centrality and Proximity on Employee Perceptions." *Administrative Science Quarterly* 38:277–303.

Janssen, M., and W. Jager. 2001. "Fashions, Habits and Changing Preferences: Simulation of Psychological Factors Affecting Market Dynamics." *Journal of Economic Psychology* 22:745–772.

Keeney, R., and H. Reiffa. 1976. "Decision with Multiple Objectives: Preferences and Value Trade-offs. New York: Wiley.

Lansing, J.S. 2003. "Complex Adaptive Systems." *Annual Review Anthropology* 32:183–204.

LeBaron, B. 2000. "Agent-Based Computational Finance: Suggested Readings and Early Research." *Journal of Economic Dynamics and Control* 24:679–702.

LeBaron, B. 2001a. "A Builder's Guide to Agent-Based Financial Markets." *Quantitative Finance* 1:254–261.

LeBaron, B. 2001b. "Empirical Regularities from Interacting Long and Short Memory Investors in an Agent-Based Financial Market." *IEEE Transactions on Evolutionary Computation* 5:442–455.

Levin, S. 1998. "Ecosystems and the Biosphere as Complex Adaptive Systems." *Ecosystems* 1:431–436.

Levin, S. 2003. "Complex Adaptive Systems: Exploring the Known, the Unknown and the Unknowable." *Bulletin of the American Mathematical Society* (new series) 40:3–19.

Luce, M. F., J. R. Bettman, and J. W. Payne. 1997. "Choice Processing in Emotionally Difficult Decisions." *Journal of Experimental Psychology: Learning, Memory, and Cognition* 23:384–405.

Olsson, P., C. Folke, and T. Hahn. 2004. "Social-Ecological Transformation for Ecosystem Management: The Development of Adaptive Co-management of a Wetland Landscape in Southern Sweden." *Ecology and Society* 9:2.

Ostrom, E. 1990. *Governing the Commons: The Evolution of Institutions for Collective Action.* Cambridge: Cambridge University Press.

Ostrom, E. 2005. *Understanding Institutional Diversity.* Princeton: Princeton University Press.

Pauly, P. 1995. "Anecdotes and the Shifting Baseline Syndrome of Fisheries." *Trends in Ecology and Evolution* 107:430.

Payne, J. W. 1976. "Task Complexity and Contingent Processing in Decision Making: An Information Search and Protocol Analysis." *Organizational Behavior and Human Performance* 16:366–387.

Payne, J. W., J. R. Bettman, and E. J. Johnson. 1992. "Behavioral Decision Research: A Constructive Processing Perspective." *Annual Review of Psychology* 43:87–131.

Redman, C. 2005. "Resilience Theory in Archaeology." *American Anthropologist* 107:70–77.

Simons, H. A. 1956. "Rational Choice and the Structure of Environments." *Psychological Review* 63:129–138.

Slovic, P. 1995. "The Construction of Preference." *American Psychologist* 50: 364–371.

Solow, R. 1974. "Intergenerational Equity and Exhaustible Resources." In *Review of Economic Studies: Symposium on the Economics of Exhaustible Resources,* 41:29–45, Edinburgh. Longman.

Stiglitz, J. 1974. "Growth with Exhaustible Natural Resources: Optimal Growth Paths." In *Review of Economic Studies: Symposium on the Economics of Exhaustible Resources,* 41:123–137. Edinburgh: Longman.

Walker, B., S. Carpenter, J. Anderies, N. Abel, G. S. Cumming, M. Janssen, L. Lebel, J. Norberg, G. D. Peterson, and R. Pritchard. 2002. "Resilience Management in Social-Ecological Systems: A Working Hypothesis for a Participatory Approach. *Conservation Ecology* 6:14. http://www.consecol.org/vol6/iss1/art14/

Walker, B., C. Holling, S. Carpenter, and A. Kinzig. 2004. "Resilience, Adaptability, and Transformability." *Ecology and Society*, vol. 9. http://www.ecologyand society.org/ vol9.iss2.

Walters, C. 1986. *Adaptive Management of Renewable Resources.* New York: Macmillan.

Walters, C. 1997. "Challenges in Adaptive Management of Riparian and Coastal Ecosystems." *Ecology and Society* vol. 1. http://www.consecol.org/vol1/iss2/art1.

Weisbrod, B.A. 1964. "Collective Consumption Services of Individual Consumption Goods." *Quarterly Journal of Economics* 78:471–77.

Wilson, J. A., J. M. Acheson, M. Metcalfe, and P. Kleban. 1994. "Chaos, Complexity, and Community Management of Fisheries." *Marine Policy* 18:291–305.

6

REGIME SHIFTS, ENVIRONMENTAL SIGNALS, UNCERTAINTY, AND POLICY CHOICE

William A. Brock, Stephen R. Carpenter, and Marten Scheffer

REGIME SHIFTS, substantial reorganizations of complex systems with prolonged consequences, have been described for many natural and social systems relevant to environmental science (Steele 1998; Scheffer et al. 2001; National Research Council 2002; Carpenter 2003; Brock 2004). In environmental policy regime shifts raise the prospect that incremental stresses may evoke large, unexpected changes in ecosystem services and human livelihoods. There is considerable interest in understanding regime shifts as well as in developing early warning indicators of impending regime shifts. This chapter addresses these issues using minimal models. We focus particularly on the problem of early warning indicators and the related issue of using data to discipline the range of model uncertainties in practical policy making.

A major goal of the chapter is to contribute to building a theory of burden of proof in contentious debates that involve environmental regime shifts. A prominent example is human-induced climate change and what, if anything, should be done about it. In this case inference is difficult because there is only one system (the global climate system) undergoing unprecedented changes. There is no opportunity to learn from replicates; the relevance of historical experience is debatable. By contrast, we consider the case of ecosystem management in which a number of similar systems are to be managed. Water quality of lakes and reservoirs provide a well-studied example (Scheffer 1998; Carpenter, Ludwig, and Brock 1999; Ludwig, Carpenter, and Brock 2003; Carpenter 2003). In this case information obtained from changes in similar systems may greatly reduce uncertainties in managing a particular system of interest.

A secondary goal is to relate the primary part of this chapter to recent work in social science that uses threshold and tipping point models to try to understand the causes of rapid shifts in social phenomena ranging from relatively trivial things such as fads, fashions, and the like to dynamics of public opinion on critical issues such as global climate change.

The first section sketches a minimal modeling framework that will be used throughout the chapter. The model represents a two-level hierarchy of dynamics where the fast dynamics is a stochastic differential equation and the slow dynamics of a bifurcation parameter is deterministic. The second section discusses basic identification problems in using data to infer what dynamical process is generating the system behavior. This section also discusses model uncertainty and offers some suggestions on how to deal with it. The third section contains a brief discussion of related issues in social science. The fourth section contains a brief application to lake systems and their management.

MODELS OF REGIME SHIFTS

Consider the following stochastic differential equation (SDE) system adapted from Berglund and Gentz (2002a, b), and Kleinen, Held, and Petschel-Held (2003),

(2.1a) $dx/dr = f(x,a) + s\, dW/dr$, $x(0)$, given,

(2.1b) $da/dr = e$, $0 < e \ll 1$, $a(0)$ given $e(0) = 0$

where x is a one-dimensional system state variable, a is a one-dimensional slow-moving parameter, dW is a Wiener process (i.e., dW is uncorrelated over time and dW is normally distributed with mean zero and variance dt), and s measures the standard deviation of the disturbances dW. Here f is called the instantaneous mean and s^2 is called the instantaneous variance, respectively. Since $a = er$, for precise work it is standard in the mathematical literature to follow Berglund and Gentz (2002a, b) and rescale time by putting slow time $t = er$. This results in the SDE

(2.1'a) $dx(t) = (1/e)f(x(t),t)dt + (s/e^{0.5})dW(t)$.

We proceed informally here by working with (2.1a) for each fixed value of parameter a and examining what happens as we step through values of parameter a, but remind the reader that much of the mathematically precise literature on slow time/fast time systems works with (2.1'a).

This framework can be generalized to the case where x is an n-dimensional vector and a is an m-dimensional vector, but we shall work with scalar cases here. The SDE (2.1) will always be an Ito SDE here, but the Stratonovich case can also be treated. See Horsthemke and Lefever (1984) for treatment of both cases.

If there is a function $F(x,a)$ such that

(2.2a) $dF(x,a)/dx = f(x,a)$,

then the system (2.1a) is a simple hills climbing system for $s = 0$ and parameter vector a fixed. For example, for an evolutionary dynamical system, one can think of $F(x,a)$ as a "fitness landscape" and a as a slow-moving bifurcation parameter where (2.1a) is the "fast" dynamics and (2.1b) is the "slow" dynamics. In the scalar case considered here, there is always such an $F(x,a)$ for each fixed a. One constructs $F(x,a)$ by integrating $f(z,a)$ over z from $z = 0$ to $z = x$.

A solution $\{X(t,x(0),a(0))\ A(t,x(0),a(0))\}$, given initial conditions, $x(0)$, $a(0)$ of (2.1) is a stochastic process. When $e = 0$, i.e., a is fixed, we shorten the notation to $\{X(t,x(0);a)\}$. Keep a fixed for now. The steady state density, $P(X = x;a)$ plays the same role in the stochastic case as does a steady state of the deterministic system when $s = 0$. We write $P(X = x;a)$ for the precise expression, $P\{X \text{ in } [x,x+dx];a\} = P(X = a;a)dx + o(dx)$, where $o(dx)/dx \rightarrow 0$ as $dx \rightarrow 0$. $P(X = x;a)$ denotes the density function of X, which depends upon the "shift" parameter a. It is well known that under regularity conditions on F (Bhattacharya and Majumdar 1980; Horsthemke and Lefever 1984), we have

(2.2b) $P\{X = x;a\} = \exp((2/s^2)\ F(x,a))/Z$ where

Z is a normalization factor so that $P\{X = x;a\}$ integrates to one. There is a vector version of (2.2b) (Bhattacharya and Majumdar 1980). See Bhattacharya and Majumdar (1980) for a discussion of existence and uniqueness of invariant measures for multivariate diffusions as well as a careful discussion of the regularity conditions needed. Their treatment does not require the existence of a "potential function" $F(x,a)$ that is highly restrictive because, at the minimum, this requires symmetry of the cross-partial derivatives of $f(x,a)$ in x as well as the more modest requirement of connectedness of the domain of $f(x,a)$ in x for each value of a.

Regularity conditions used by Bhattacharya and Majumdar include connectedness of the domain and thrice differentiability of the instantaneous mean function $f(x,a)$ and instantaneous standard deviation function (s in our case, $s(x,a)$ in Bhattacharya and Majumdar's case). Existence and uniqueness of invariant measures do not require the restrictive sufficient conditions for existence of a "potential function" $F(x,a)$ such that $F'(x,a) = f(x,a)$. Hence it should be possible to generalize some of the results discussed in this chapter to general $f(x,a)$. More will be said about this below.

Indeed, under modest regularity conditions $F(x,a)$ always exists for the case where x is one-dimensional. Just set $F(x,a)$ equal to the integral of $f(z,a)$ between x_0 and x for a fixed x_0. One can apply Bhattacharya and Majumdar to locate sufficient conditions for existence and uniqueness of an invariant measure for general $f(x,a)$ when x is one dimensional. Berglund and Gentz (2002a, b) provide the sufficient conditions for the stochastic bifurcation theory that we discuss below.

The disturbances of dx/dt cause transient changes in x. These changes are inversely proportional to the slope of $f(x,a)$ near the stable equilibrium point (figure 6.1). In rough terms, then, the variance of x will be larger the smaller the slope of $f(x,a)$ near the stable equilibrium point. If the system is near a stable point and slow changes in a are causing the slope of $f(x,a)$ to decrease, then the variance of x should increase (figure 6.2).

Others have noted that changes in variance over time may provide a clue to impending bifurcation in environmental systems (Kleinen, Held, and Petschel-Held 2003; see box 6.1). An approach used in economics depends on the fact that if one has a noise-free time series on x over any interval $(t,t + dt)$, then s can be estimated to any degree of accuracy within $(t,t + dt)$ by sampling at higher and higher frequencies within $(t,t + dt)$. This is called continuous record asymptotics and it plays a big role in the estimation of conditional variances in finance (Foster and Nelson 1996; Campbell, Lo, and MacKinlay 1997).

If one has structural information on how s relates to other features of the system, e.g., $f(x,a)$, one can exploit this information using continous record asymptotics and get better estimates of $f(x,a)$. There are some applications

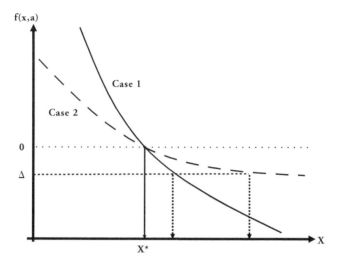

FIGURE 6.1. Plot of $f(x,a)$ versus x, illustrating the effect of the slope of $f(x,a)$ near steady state (x^*) on the value of x after a small perturbation Δ. In case 1 the slope of $f(x,a)$ near x^* is relatively large, and the change in x immediately following the perturbation is relatively small. In case 2 the slope of $f(x,a)$ near x^* is smaller, and the change in x following the perturbation is relatively large.

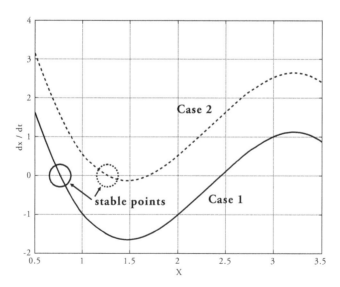

FIGURE 6.2. Plot of dx/dt versus x for a system with multiple stable points, illustrating how the slope of $f(x,a)$ near a stable point changes as the system approaches a bifurcation. Case 1: system relatively far from bifurcation. Case 2: system closer to bifurcation.

BOX 6.1

GLOBAL CLIMATE: CHANGING VARIANCE AS A CLUE TO BIFURCATION?

There is considerable interest in the possibility that gradual increases in the emission of greenhouse gases could lead to large changes in the climate system (National Research Council 2002). This general problem was considered in an abstract way by Kleinen, Held, and Petschel Held (2003). Consider their model

(B1.1) $dx = (x^2 - x + a)dt + s\,dW := f(x,a)dt + s\,dW$,

$F(x,a) = (1/3)x^3 - (1/2)x^2 + ax$.

It is easy to see that steady states for the deterministic system are given by

(B1.2) $x_1 = \frac{1}{2} - [(\frac{1}{4}) - a]^{\frac{1}{2}}, x_2 = \frac{1}{2} + [(\frac{1}{4}) - a]^{\frac{1}{2}}$,

where, for $a < a_c := \frac{1}{4}$, x_1 is locally stable, x_2 is locally unstable, and a saddle node bifurcation takes place at $a = \frac{1}{4}$. The goal of environmental monitoring is to provide a warning when the system approaches the bifurcation.

Kleinen, Held, and Petschel-Held (2003) propose to estimate the distance da from bifurcation by looking at the spectrum of x and study how it shifts as da slowly changes. If we linearize the Kleinen, Held, and Petschel-Held (2003) system at the locally asymptotically stable steady state x_1, we find that for

(B1.3) $B = 0, A = 2x_1 - 1 = -[(\frac{1}{4}) - a]^{\frac{1}{2}}$,

we have

(B1.4) $Ez^* = 0, V(z^*) = s^2 / [(\frac{1}{4}) - a]^{\frac{1}{2}}$,

for $a < \frac{1}{4}$.

Kleinen, Held, and Petschel-Held (2003) suggest estimating the spectrum generated by (B.1) over segments of time $[T,T+N]$ where N is large enough so that the steady state distribution is a good approximation, and monitoring the change in shape of the spectrum. A more direct method is to use continuous record asymptotics. As Kleinen, Held, and Petschel-Held point out, $(\frac{1}{4}) - a$ is the distance from bifurcation for $a < \frac{1}{4}$ and we wish to use time series records to estimate this quantity. One way to do it is to use a rolling window estimate of variance that uses data from t to $t+N$. Omitting data before t prevents the estimator from being overwhelmed by history and thereby becoming insensitive to more recent data which contains information on more recent values of a. Once we have an estimator of $V = = s^2 / [(\frac{1}{4}) - a]^{\frac{1}{2}}$, we can estimate the trend in the distance from bifurcation $(\frac{1}{4}) - a$ because s is assumed constant.

in social science where $f(x,a)$ depends upon s, e.g., one component of the vector a is s itself. Although there is a limit on how much continuous record asymptotics can improve the estimate of the conditional mean (unlike the improvements available for estimation of the conditional variance), obviously improvement can be had for cases where f explicitly depends upon s.

Environmental systems have some important complications, such as observation errors in the time series of x. Also, some environmental data are most sensibly collected at discrete time intervals (e.g., daily or annually), and this may limit the range of frequencies that can be studied. Nevertheless, the method of continuous record asymptotics offers an interesting approach for empirical study of environmental systems governed by equations like 2.1.

Here are some examples of systems that can be put into the framework (2.1) besides the ones mentioned by Berglund and Gentz (2002a, b), and Kleinen, Held, and Petschel-Held (2003).

Consider the social system

$$(2.3)\ dx/dt = \tanh(h+Jx)-x + s\,dW/dt,\ a := (h,J),$$

studied, for the deterministic case $s = 0$, by Brock and Durlauf (2001a, b), Scheffer, Westley, and Brock (2000, 2003), and Brock (2004). In this case $x(t)$ is the difference, at date t, between the fraction of a population that chooses a choice -1 (e.g., do not regulate emissions of greenhouse pollutants) and the fraction that chooses a choice $+1$ (e.g., do regulate emissions of greenhouse pollutants) where h stands for the utility difference between the two choices, J is a measure of the strength of social interactions or peer effects, the hyperbolic tangent function, $\tanh$, emerges from a discrete choice model, and dW/dt represents outside shocks to the system (e.g., climate events that influence the public's attitudes toward climate change.). System (2.3) will be discussed in the fourth section of this chapter.

Brock (2004) reviews work on "punctuated policy changes" in economics, political science, and other areas of social sciences using a system like (2.3) as a central organizing expository vehicle. We will go beyond Brock in this article by beginning a study of the impact of noise-induced transitions that adapts work of Berglund and Gentz (2002a, b), Kleinen, Held, and Petschel-Held (2003), and, especially, Horsthemke and Lefever (1984).

For another example consider the dynamical system of the state of a lake studied by Scheffer (1998), Carpenter, Ludwig, and Brock (1999), Ludwig,

Carpenter, and Brock (2003), and Carpenter (2003). A minimal model of this system is

(2.4) $dx/dt = c + f(x,a) + s\, dW/dt$

where x is phosphorus sequestered in algae, c is loading from outside activities such as agriculture, developments, etc., a is a slow-moving parameter (e.g., sedimented phosphorus), and dW/dt is outside shocks to the system. Here the payoff is $U(x,c)$, which decreases in x and increases in c. The control design problem is to design a control $c = C(x)$ to optimize the payoff U. Carpenter, Ludwig, and Brock (1999) and Ludwig, Carpenter, and Brock (2003) study the problem of designing a control sequence to optimize the discounted sum of payoffs.

It is clear that a large class of examples can be subsumed under the framework of (2.1). Suppose we have measurements of x and, maybe, but not always, a, which may be corrupted with measurement noise. We are concerned here with the following questions:

1. If we have a time series of observations on our measurements, what patterns in such a time series would give us an early warning signal of an impending bifurcation?
2. If there is a payoff $U(x,a)$ of the state of the system, how should we design a control $C(x,a)$ to optimize the payoff?
3. If the true system is $\{F^*(x^*,a^*),s^*;U^*(x^*,a^*)\}$ but we specify it as $\{F(x,a),s; U(x,a)\}$ how should we design the control $C(x,a)$ to be robust against such misspecification? How can we use our observational time series data on our measurements of x and a to discipline the set of specifications we must consider in order to make our control more robust?
4. Suppose instead of just one system under observation we have a collection of systems $\{F(x(i),a(i),i),s(i),\ U(x(i),a(i),i),\ i = 1,2, \dots , I\}$ under observation. How may we use information on a subset of such systems that have passed through a regime change to forecast impending regime change on other systems that have not passed through a regime change?
5. How might one use data to estimate features of the system (2.1)?

We do not pretend to answer all these questions in this chapter. Instead we consider two polar cases. The first is the global climate system, where there is only one system under scrutiny and our main problem is to glean evidence of impending regime change from patterns in the time series of observations (box 6.1). This is extremely difficult. The second leading case is one in which there exists a collection of systems 2.1 with observations on each, such as the problem of managing water quality for a collection of lakes. This second problem is more tractable because of the multiplicity of systems available for study.

A GENERAL IDENTIFICATION PROBLEM AND POLICY ACTION

We extend the mathematical models sketched above to put forth a simple framework that we hope sheds light on how one might usefully deal with heated debates in policy making. There are two basic issues raised by this problem. First there is a basic empirical identification problem in using time series data or any data to adduce evidence for or against alternative stable states. Second, there is model uncertainty. We discuss the identification problem first.

An important problem of model identification is the separation of spurious evidence, caused by dynamics of unobserved state variables, from true evidence of alternative stable states. This problem looms especially large in empirical work that attempts to use data to separate endogenous dynamics from exogenous dynamics, which turns out to be very difficult in social science (e.g. Brock and Durlauf 2001a, b). There is no reason to suspect the problem would be any easier in other sciences.

Here we will suggest some approaches for addressing this problem. To focus discussion, we consider a highly simplified model of the identification problem (box 6.2)

First we can attempt to use continuous record asymptotics to build estimators of $s_1(x_1,x_2)$ based upon data on x_1 even though we can not observe x_2, if x_2 moves slowly enough relative to x_1. The intuition is that if x_2 is constant over an interval $(t,t + dt)$ then the continuous record estimator in (3.1) below, call it $S_1^2(dt,n)$, converges to the true value of $s_1(x_1,x_2)$ on $(t,t + dt)$. This idea can be made more precise by using the work of Foster and Nelson (1996). Construct n equally spaced subintervals from $j = 1$ to $j = n$ of the interval $(t,t + dt)$ and construct the estimator

$$(3.1)\ S_1^2(dt,n) := (1/dt)\ \Sigma_j\{[x_1(t\text{-}(j\text{-}1)dt/n) - x_1(t - j\ dt/n)]^2\}.$$

BOX 6.2

MINIMAL MODEL OF THE IDENTIFICATION PROBLEM FOR REGIME SHIFT

A highly simplified heuristic model of the identification problem for regime shift of a pollutant-driven regime shift such as climate change is presented here.

(B2.1) $dx_1/dt - b c + f_1(x_1,a_1,x_2) + s_1 dW_1, s_1 = s_1(x_1,x_2)$

(B2.2) $da_1/dt = e_1 A_1(x_1,x_2)$

(B2.3) $dx_2/dt = e_2 f_2(x_1,x_2,a_2) + s_2 dW_2, s_2=s_2(x_1,x_2,e_2)$

(B2.4) $da_2/dt = e_3 A_2(x_1,x_2)$

(B2.5) $U(c,x_1) = c-(\frac{1}{2})Px_1^2,$

where x_1 is the state of the climate with higher x_1 corresponding to worse climate and U is the payoff to humans of human consumptive activities and the state of the climate x_1. The slope b represents pollutant emissions per unit of consumptive activity. The scalar x_2 and the dynamics (B2.3) represent other factors that shift the climate system and make it difficult for an analyst to sort out whether the climate is shifting because b>0 or whether it is shifting because of these other factors.

There are four scales of time with (B2.1) having the reference scale of one. (B2.2) has scale e_1, (B2.3) the scale e_2, and (B2.4) the scale e_3. In the Kleinen, Held, and Petschel-Held (2003) example the "confounding" variable x_2 was not present and e_1 was a slow scale, i.e. $0 < e_1 \ll 1$. This is the situation analyzed by Berglund and Gentz (2002). The most difficult identification case is where $e_3 = e_1 := e, e_2=1$.

The dynamics of x_2 impacting the dynamics of x_1 make it difficult to adduce evidence for or against the presence of alternative stable states in the dynamics of x_1. In order to bring out these empirical issues clearly we assume that x_1 is observable but x_2 is not. We allow the standard deviation functions to depend upon x_1 and x_2.

We assume for each fixed value of a that there is a landscape function $H(x_1,x_2,a)$, sometimes called a "potential function," such that

(B2.6) $f_1 = \partial H/\partial x_1, f_2 = \partial H/\partial x_2,$

Of course there is no reason why the symmetry conditions for cross partials needed for existence of a landscape function should hold. But we focus on this case so we can use the useful expository devices of a landscape diagram (for the metaphor of climbing up) and cup and ball diagram (for the metaphor of seeking the lowest point in a valley). Under the assumption of existence of a landscape, if one is more comfortable with cup and ball diagrams, the deterministic system with c−0 moves to local minima of the function $V(x_1,x_2,a) := -H(x_1,x_2,a)$.

Foster and Nelson (1996) show that, under modest regularity conditions, the increments in (3.1) are approximately independently and identically distributed normal with mean zero and variance $s_1(x_1,x_2)^2$. Hence, if we are able to sample within $(t, t + dt)$ at any frequency we like, we can essentially assume that s_1^2 is observable as well as x_1.

One could also consider empirical methods that have been used in areas like finance to estimate diffusion processes of the form

(3.2) $dx = f(x,a)dt + s(x,a)\, dW$,

Estimators of systems like (3.2) can be constructed using time series data on x over a finite interval t in $[0,T]$ (e.g., Hansen and Scheinkman 1995; Conley et al. 1997 and their references). We explain the basic idea here. Fix a over $[0,T]$. Assume conditions on f and s so that there is a unique stationary density $P\{X = x;a\}$. See Bhattacharya and Majumdar (1980), Horsthemke and Lefever (1984), Hansen and Scheinkman (1995), Conley et al. (1997) for precise conditions.

Now let $H(x)$ be any smooth function of x. On the stationary distribution EH is constant through time and, hence, its time derivative is zero. Therefore we compute by expanding $H(X(t + dt)) = H(x+X(t + dt)-x)$ in a Taylor series around x,

(3.3) $0 = E[E\{H(X(t + dt))|X(t) = x\}-H(x)\}/dt]-> E[H'f + (1/2)H''s^2]$,
$dt-> 0$.

Since the expectation E can be replaced with a sample average, and the function H was arbitrary, we can construct an infinite number of moment conditions using (3.3). Conley et al. (1997) show how to choose a collection of H's to do the best job of estimating f under regularity conditions by adapting the generalized method of moments (GMM), following the theory of Hansen and Scheinkman (1995). This method might be adapted to give us an early warning signal of impending bifurcation in $f(x,a)$ by constructing good estimators of f over moving intervals $[T,T + N]$.

The GMM can also be adapted to address the problem of observation error. Consider the discrete-time model of Carpenter (2003), which uses Bayesian methods to estimate both linear and nonlinear models to detect possible regime shifts in the presence of measurement error. We discuss here how GMM might be used. Consider the system

(3.4) $x(t + 1) = f(x(t),b) + e(t + 1)$, $y(t) = x(t) + m(t)$

where $\{e\}$, $\{m\}$ are mutually independent, identically distributed mean zero finite variance processes. Here $\{y(t)\}$ is observed but $\{x(t)\}$ is not, and the task is to estimate the parameter vector b. Consider a test function $H(y)$ and examine the quantity

(3.5) $H(y(t)) [y(t+1)-m(t+1)-f(y(t)-m(t),b)] = H(y(t)) e(t+1)$.

Under our assumptions of mutual independence and independence over time of $\{e\}$, and $\{m\}$, we have

(3.6) $0 = E\{H(y(t)) e(t+1)\} = E\{H(y(t) [y(t+1)-((f(y(t)-m,b) g_m(m) dm]\}$.

Hence, if we assume a known density g_m for the measurement error m, e.g., normal with mean zero and finite variance, we have a GMM system. Hence we can apply standard GMM under regularity conditions following Hansen (1982) and choose a set of "test functions" $\{H\}$ that give us good estimates of b. It would be interesting to compare the performance of GMM methods with the methods used by Carpenter (2003).

The problem of model uncertainty is fundamental to science-based dis-agreements about environmental policy (Brock, Durlauf, and West 2003; Carpenter 2003). Consider climate change as an example (box 6.2). Suppose there is only one stable state in the observed variable $x1$, which is the ob-served state of the climate. Further, suppose the $x2$ variable (an unobserved dynamic variable that impacts the climate) operates on a slow scale of time in such a way that the $x1$ dynamics generates time series data that resembles a system with alternative stable states. Suppose there are two plausible types of social actions on management of potential climate change. Type 1 is cau-tious, with low b and low emission of greenhouse gas, type 2 is the opposite. Hence the policy maker faces four possible outcomes, depending upon her choice of actions as well as the resolution of the state of model uncertainty represented by the $x2$ variable. The most benign outcome is type 1 policy combined with the "good" value of $x2$; the worst outcome is the type 2 policy combined with the "bad" value of $x2$. The two other pairs are intermediate.

Similar model uncertainty problems were addressed by Brock, Durlauf, and West (2003) and Carpenter (2003). The former addresses model un-certainties related to monetary and economic growth policy, the latter addresses model uncertainties in fishery management. Both papers employ Bayesian model averaging (BMA) methods. Brock, Durlauf, and West (2003) use methods from the robustness literature and literature on ambiguity aversion to frame a data-based approach to policy action. They conduct two empirical exercises, one in monetary policy, the other in economic growth policy, using linear models. Carpenter (2003) applies BMA to both linear and nonlinear models. The regime shift problem could be addressed using a nonlinear version of Brock, Durlauf, and West (2003) using models similar to those of Carpenter (2003) and supplemented with methods from robust analysis and methods to deal with ambiguity aversion. Our discussion of the model of box 6.2 suggests the use of an action dispersion plot with two choices of action and two choices (by nature) of the state of model uncertainty. Brock, Durlauf, and West (2004) argue for the presentation of empirical action dispersion plots, in advisement of policy makers by scientists, and present such plots for monetary policy. Such plots are useful in environmental policy disputes for exactly the same reasons.

We believe our review of new literature on estimation techniques and dealing with model uncertainty should be useful for both scientists who must report to policy makers and to policy makers who must make demands on scientists to present their results in an understandable manner to the policy makers together with an honest reporting of the true level of uncertainty. However, there is tremendous need for empirical exploration of concrete, data-driven examples in environmental science. We turn now to a brief discussion of related methods in social science.

SOCIAL SYSTEMS AND COUPLED NATURAL AND SOCIAL SYSTEMS

Wood and Doan (2003) build a regime change theory to argue that "whenever there is a preexisting condition that many find privately costly, but with widespread public acceptance, the system is ripe for change." They conducted an empirical exercise on public attitudes to sexual harassment and found evidence consistent with a regime shift around the time of the Clarence Thomas hearings. Brock (2004) provides a broad review of similar types of regime change in social science, many of which could be addressed by the methods described above.

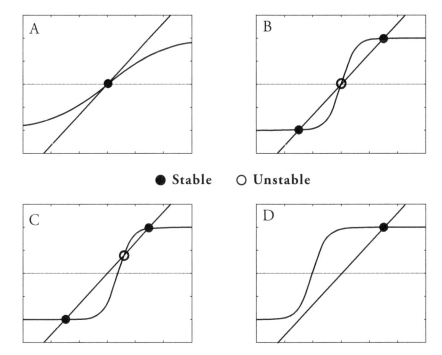

● **Stable**   ○ **Unstable**

FIGURE 6.3. (A) Split population ($h = 1$) with low social interactions (low J). There is a single equilibrium. (B) Split population with high social interactions (high J). There are three equilibria, two of which are stable. (C) Biased population ($h < 0$) with high social interactions. The unstable threshold has moved to the right. (D) News spreads among a population with high social interaction, causing h to rise above zero. Two of the equilibria vanish, leaving only one stable equilibrium.

A main interest in social interactions studies (modeled by [2.3] above) is the dynamics of polarization due to non-negative social interactions effects $J > 0$. Since $f(x,a) := \tanh(h + Jx) - x$ is scalar, it is trivial to find $F(x,a)$ such that $dF(x,a)/dx = f(x,a)$. It is easy to show that system (2.3) has only one stable state for $J < 1$ and that if J slowly increases, a pitchfork bifurcation appears if J becomes positive and large enough (compare panels A and B of figure 6.3). We first discuss the deterministic case $s = 0$, then we discuss what happens when s becomes positive.

First consider the deterministic case $s = 0$ with $h = 0$ and note that the only stable state is $x = 0$ for small J (figure 6.3a). As J increases beyond unity, two new stable states appear close to $x = 0$ and depart further away from $x = 0$ (which becomes unstable) as J continues to increase (figure 6.3b).

Another bifurcation, called a saddle node bifurcation can be produced at, for example, the negative stable state for $J > 1$. Do this by slowly increasing h from its initial value of zero. As h is slowly increased, a critical value of h will be reached where the negative stable state and the steady state at zero come together and both vanish as h continues to increase, leaving only the positive stable state (figure 6.3d).

A good example is to think of the state $x = 0$ as representing an evenly divided electorate where half are for an incumbent candidate and half are against. The quantity J measures the cost of deviating from the average view of the whole population. Think of it as a measure of conformity pressure. As J increases beyond unity, two new stable states appear, one "anti-incumbent," the other "pro-incumbent," while the zero stable state loses its stability (panel A to panel B, figure 6.3). As J becomes very large, the two new stable states are pushed close to -1 and $+1$ (figure 6.3b).

Now suppose that h has been negative in the past, i.e., the incumbent has been preferred to alternatives (figure 6.3c). Also suppose that social pressure to conform to the majority opinion is very high so that J is very large. So the system is at the negative stable state and it is close to -1. But as news appears that is unfavorable to the incumbent, h slowly rises. In the deterministic case $s = 0$, the system remains stuck at the pro-incumbent position but becomes unstuck when h becomes large enough that the negative stable state vanishes (panel C to panel D of figure 6.3). At this point the system moves rapidly toward the positive stable state. This is an example of what is sometimes called a "macropunctuated change" in political science.

Now let us see what happens when s is positive. Let $a := (h,J)$ and let $F(x,a)$ denote the integral of $\tanh(h + Jz) \cdot z$ up to x. Then the steady state density of x is given by

(4.1) $Pr\{X = x; a\} = \exp[(2/s^2) F(x,a)] / Z$,

where Z is a normalization factor so (4.1) integrates to one. Obviously, as s converges to zero (4.1) and spikes at the global maximum of F, and the global maximum of F shifts abruptly for the case $J > 1$ from a large in absolute value negative x to a positive x as h passes through zero. In other words, a slight tilt of h against the incumbent, i.e., $h > 0$, can cause a highly polarized electorate (a high value of J) to cluster at a new position, i.e., a large and positive (almost symmetrically opposite) value of x.

The reader may ask what happened to the hysteretic stickiness that was present in the case $s = 0$. It is easiest for many to think in terms of a cup and ball metaphor. Think of $-F$ as a potential by writing $dx = fdt + sdW$ as $dx = -(-dF/dx)dt + sdW$. Then we now have a "rattling" ball in a cup. If s is small, a ball lying in a negative well of a two-welled potential will have a hard time getting out, although it will get out with probability one. Once the ball drops into a deeper well, it is relatively harder to get out, but it will with probability one escape the deeper well and fall back into the more shallow one. As s becomes smaller, the relative probabilities shift more toward staying in the deeper well and in the limit as s goes to zero, the ball will end up in the deeper well. This idea is closely related to the method of simulated annealing in numerical optimization (Kirkpatrick, Gelatt, and Vecchi 1983).

Berglund and Gentz (2002a) point out that, in the "small noise limit," the expected time between transitions can be approximated by Kramer's formula, which is proportional to $\exp[(2/s^2)G]$ where G is the barrier height between the two wells. Hence when G is high and s is low we expect it to take a long time to move from a pro-incumbent position, i.e., the negative x stable state, to an anti-incumbent position, i.e., the positive x stable state. An increase in J increases G. This is a stochastic analog of hysteresis in the deterministic case. Berglund and Gentz (2002a) and Norberg et al. (2001) study stochastic systems with periodic forcing. These studies give us some insight into what to expect when the system is forced by a slow moving, but nonperiodic and possibly random force. Turn now to an ensemble type of social interactions model where the interaction between the increasing size of the system and the outside shocking process produces new nonhysteretic behavior as the size of the system approaches infinity.

There is an interesting model building on Dawson (1983) that helps us understand why tipping points occur in Brock (2004) as well as in Wood and Doan (2003). It does not appear in those papers. We exposit it here.

Consider the stochastic differential equation

(4.2) $dx = (-x^3 + a\,x)dt + s\,dW.$

As s goes to zero the mass of the stationary density, $P\{X = x;a\}$ clumps onto a Dirac delta distribution spike at

(4.3) $x^*(a) = \operatorname{argmax}\{-(\tfrac{1}{4})x^4 + a\,x^2/2\}.$

For $a = 1$, $x^*(1)$ is -1 and $+1$ and as a becomes greater than one, the global maximum is the larger local maximum. This sets the stage for the coupled system, where $dWdW' = Idt$,

(4.4) $dx_i = [-x_i + a\,x_i]dt + s\,dW_i - J(x_i - xbar)$, $xbar := (1/N)\Sigma_i\{x_i\}$,

where the Sum runs from $i = 1, 2, \ldots, N$. Note that dW is an Nx1 vector here. Put $x = (x_1, \ldots, x_N)$. Following Bhattacharya and Majumdar (1980), we find the invariant measure $P(X = x) = \exp(bU(x))/Z$, $b := 2/s^2$, by constructing $U(x)$ such that

(4.5) $dx = dU(x)/dx + s\,dW$

for an appropriate "potential function" $U(x)$. It is easy to check that the cross-partial symmetry conditions needed for existence of U are satisfied. One may check that U given by

(4.6) $U(x) = \Sigma_i[u(x_i) - Jx_i^2/2] + (J/2)[xbar]^2$, $u(_i j) := -(\tfrac{1}{4})x_i^4 + a\,x_i^2/2$

is appropriate.

Dawson (1983) studies the limiting behavior of a system very close to this one. It is basically the same as systems of coupled oscillators in statistical physics. The thrust of this type of work is to locate sufficient conditions for the average $\Sigma\{x_i\}/N$ to converge in an appropriate probabilistic sense to the global maximum of some deterministic function, call it S. Brock and Durlauf (2001a, b) show that there is a close relationship between this literature and that on discrete choice modeling in econometrics. The deterministic function, S, in the discrete choice case turns out to be a measure of limiting expectation of maximal social welfare per capita and the average, $\Sigma\{x_i\}/N$, converges to the global maximum of that function. It turns out that small changes in the individual payoffs coupled with large enough values of the conformity index J cause large movements in the limiting value of the average behavior as N becomes large. Brock (2004) attempts to explain this type of behavior in intuitive language and apply it to predicting apparent phase transitions in social systems. The main finding is this. The hysteretic sticky movement across different stable states as payoffs slowly change gets replaced by a nonhysteretic jump to the global maximum of a function S, wherever that global maximum may move. In other words, the social system becomes more smoothly adaptive to changing conditions as s increases. Turn now to a brief discussion of dependence of our methods on the assumption that a "potential" function $F(x,a)$ exists such that $F'(x,a) = f(x,a)$ for each a.

If one sets the shocks to zero, there is a classification theory for bifurcations for the differential equation $dx/dt = f(x,a)$ (Kuznetsov 1995). For example, for the case where parameter a is one-dimensional, there are two "primary" bifurcation types, one is when the "largest" eigenvalue of the Jacobian matrix, $f'(x^*(a),a)$ is real and passes from negative to positive as parameter a increases and the other is when the "largest" eigenvalue is complex and the pair consisting of this eigenvalue and its complex conjugate share a real part that passes from negative to positive. Here $x^*(a)$ denotes a solution of the steady state deterministic equation, $0 = f(x^*(a),a)$ for each value of parameter a.

There is also a classification theory for the next "level" of bifurcation that concerns the induced Poincaré map when a limit cycle appears. There is also a theory of "global" bifurcations, for example, the homoclinic bifurcation. All this is covered by Kuznetsov (1995) and none of it is dependent upon existence of a "potential" $F(x,a)$ such that $F'(x,a) = f(x,a)$. The corresponding theory of closed-form analytic expressions for invariant measures is not as well developed in the case of general $f(x,a)$ as it is for the "integrable" case where $F'(x,a) = f(x,a)$ for some "potential" function $F(x,a)$ that maps x-space to the real line for each value of a. The symmetry of cross partials of $f(x,a)$ plays a big role in producing analytic closed-form expressions for invariant measures.

However, one can still obtain computational results for stochastic bifurcations quite easily. This is a centerpiece of the research program of Cars Hommes and his CeNDEF group in Amsterdam. See Hommes's review (2005) for example. In Hommes's research strategy one uses bifurcation classification theory to classify the primary (and in some cases the more refined secondary bifurcations) for the analytical part of the research strategy. Then one adds the noise, sdW, for small s and turns to the computer. Using the computer one can produce the analogs of bifurcation diagrams for the small noise stochastic case and study what happens. We believe that such a study might give useful insight into observable signals of impending bifurcations as one moves parameter a slowly toward and through a bifurcation point.

If the noise is very small one can produce a local linear approximation to the solution of (2.1a) for general vector cases and use "small noise asymptotics" (e.g., Magill 1977) to produce closed form analytical expressions for objects such as the spectral density matrix of the local linear approxima- tion. This possibility suggests that it might be useful to step parameter a through a bifurcation point and study what happens to the spectral density matrix of the local linear approximation as one steps parameter a through a bifurcation point. This would be a generalization of the Kleinen, Held, and Petschel-Held (2003) approach for vector cases where additional types of bi- furcations, for example, the Hopf bifurcation, can occur. Local linearization "small noise" approximation theory does not depend upon the existence of a potential. Neither does the use of GMM-based methods to estimate $f(x,a)$ from continuous record data. It is beyond the scope of this chapter to pursue these potentially promising research avenues and generalizations further.

COLLECTIONS OF MANY SIMILAR SYSTEMS

The ecosystem regime changes of Scheffer (1998), Carpenter, Ludwig, and Brock (1999) and Carpenter (2003) are situations where a large number of similar systems can be studied, some of which have passed across thresh- olds. Although the systems are not identical replicates, they are similar enough that information from one set of systems is transferable to another set of systems. Analogous situations occur in panel data studied by social scientists.

By studying many shallow lakes, for example, it has become clear that lakes with a total phosphorus level of more than 0.1 mg l^{-1} are at risk of col- lapsing to a turbid state (Jeppesen et al. 1990; Scheffer 1998). Importantly, a single threshold level will never apply to all systems. In terms of bifur- cation theory, the bifurcation point will always be dependent on various parameters. In the case of lakes, for instance, differences in size will affect their sensitivity to collapse at increasing phosphorus levels. A recent study of 240 shallow floodplain lakes that are similar in nutrient level due to annual inundation by the Rhine showed that the probability of being in a clear state is considerably higher for smaller lakes (Van Geest et al. 2003). Thus, if one has the possibility to study many similar systems, empirical indicators for the risk of regime shifts may be obtained.

Carpenter (2003) shows how measurements from many lakes can be used to construct informative priors for Bayesian estimation of systems like 3.4. Such informative priors reduce the uncertainty of threshold estimates. In addition, the informative priors increase the posterior weight of the correct model in simulation studies that use Bayesian model averaging to compute policies under model uncertainty.

The tendency of variance to increase near thresholds has not been widely exploited by ecosystem scientists. The intuitive reason for the increase in variance can be seen by noticing that as parameter a moves from figure 6.3a to figure 6.3b very slowly, parameter a "flattens" the slope of $f(x,a)$ toward unity. This action magnifies the impact of an outside shock, sdW. This suggests that if one could follow a cross section of lakes, some nearer to bifurcation than others, it might be possible to use such data to forecast impending bifurcations.

For example, it would be interesting to explore indicator variates that can be measured at high frequency with low observation error. One possibility is the volume of anoxic water in a lake subject to eutrophication. In deep thermally stratified lakes, the volume of anoxic water is directly related to the proximity of a threshold for eutrophication (Carpenter, Ludwig, and Brock 1999; Carpenter 2003) and formation of marine hypoxic zones (Stow, Qian, and Craig 2005). Technology exists for accurate and rapid monitoring of oxygen in lakes. It may be possible to devise monitoring schemes using continuous record asymptotics, as described above, to create leading indicators of breakdowns in water quality for lakes and reservoirs (Stow, Qian, and Craig 2005).

In using indicators, one must be cautious about the identification problems described in box 6.2 and related material from the third section of this chapter. For example, variance of fish populations and their prey might be expected to increase near a depensation point, a type of threshold that occurs in population dynamics (Carpenter 2003). However, there is also evidence that maximum-sustainable-yield-type management schemes cause increases in the variance of fish populations and entire pelagic food webs through trophic cascades (Carpenter and Kitchell 1993). The ecological conditions near an MSY target can be quite different from those near a depensation point. Thus increases in the variance of fish populations may have ambiguous implications for ecosystem dynamics.

On the other hand, there may be specific subtle changes in a particular system that signal deterioration of the resilience of its current state. Such changes are not generic (such as coloring of noise) but rather unique to the type of system under study. For instance, in shallow lakes a shift to a turbid state is typically preceded by an increase of the periphyton layer covering the macrophytes and a reduction in the proportion of piscivorous fish (Meijer et al. 1994; Scheffer 1998). Knowing such clues of proximate regime shifts requires a mechanistic insight into the functioning of the system, which is more likely if one has the possibility to study many systems.

The multiplicity of lakes on the landscape offers the possibility of mosaic management (Carpenter and Brock 2004). In mosaic management different ecosystems are managed for different objectives, including deliberate experimentation to reduce uncertainty. Mosaic management is also an approach for addressing strongly divergent social goals for ecosystems (for example, lakes for water supply or recreation versus lakes for dilution of pollutants). On the other hand, mosaic management leads to the possibility of complex spatial dynamics of ecosystem users, which may create new types of thresholds on complex landscapes (Carpenter and Brock 2004).

Can early warnings of regime shifts come soon enough for people to act to avert an unwanted regime shift? In rare cases, in which numerous similar systems have been intensively studied, such as in the case of shallow lakes, empirical rough rules of thumb of where the threshold lies may sometimes be found. However, this situation seems the exception rather than the rule. Simulation studies using phosphorus recycling as an indicator of impending lake eutrophication suggest that decision makers will receive one to three years advance warning of breakdowns in lake water quality (Carpenter 2003). This lead time is not sufficient for effective action by any existing lake management system. Kleinen, Held, and Petschel-Held (2003) suggest that the time scales for detecting climate regime shifts are about the same as the time scales for effective action to prevent the regime shifts (one hundred to one thousand years). Thus it seems unlikely that an impending regime shift could be detected in time to evoke effective action, unless the management system allowed for rapid and massive response. In practice there are many difficulties with implementing early warning and rapid response systems (Sarewitz, Pielke, and Byerly 2000). More conclusive evidence for the possibility and social costs of crossing thresholds would contribute to the general problem of designing adaptive strategies for environmental regime shifts.

This chapter has centered on an identification problem and a policy design problem. The identification problem is to determine whether threshold dynamics can occur and whether a threshold is near. The identification problem is complicated by hidden variables with return times near those of key state variables, observation error, and the multiplicity of plausible models for many social-ecological systems. We suggest some approaches for addressing these challenges. Our review of the literature shows how dynamics of the variance of state variables near a threshold could provide some advance warning of impending regime shifts.

The policy design problem is to identify patterns of evidence that should prompt us to choose actions to avert unwanted and impending regime shifts. Growing evidence for environmental thresholds and regime shifts suggests that this policy design problem will become more prominent in coming decades.

This policy design problem is complicated by the presence of more than one model that gains substantial support from data and basic scientific understanding. Brock, Durlauf, and West (2003) discuss this problem in detail in the contexts of monetary policy and growth policy. Their discussion is pertinent to the regime shifts discussed here.

Bayesian model averaging is one way to report the fundamental uncertainty about which model is appropriate. A closely related approach is to report the entire posterior distribution of payoffs to the policy maker over the set of models, weighted by the posterior probability of each of these models when fitted to the available data. The policy maker, when presented with this posterior distribution, can then choose an action by whatever preferences he has. Evidence that dates back to Ellsberg suggests that decision makers include some aspects of avoidance of worst case scenarios in their preferences. In other words, they do not act like a Bayesian using Bayesian model averaging of the payoffs when they choose their optimal action. To put it another way, they might act like they put some weight on the worst case scenario and some weight on the Bayesian model average. This observation suggests the possibility of developing a theory of how much of the burden of proof each side should bear in scientific disputes.

In our context, it is especially important to discount past observations as we collect more and more current information as the system moves forward through time. Contrasting views may converge as more evidence comes in. Discounting past observations is important to avoid Bayesian posteriors being frozen by history when the system may be approaching a bifurcation point. Discounting is important because the distant past contains less information about an impending bifurcation than the recent past.

The use of techniques like continuous record asymptotics to estimate conditional variance has the potential to sharpen development of early warning indicators. The approach may also sharpen the development of theories of the burden of proof in scientific disputes in situations where impending regime change is possible and has significant implications for human welfare.

REFERENCES

We are grateful to Simon Levin, Jon Norberg, and an anonymous referee for helpful comments on the text. William A. Brock acknowledges support from the Vilas Trust. William A. Brock and Stephen R. Carpenter are grateful for NSF support.

Berglund, N., and B. Gentz. 2002a. "A Sample-Paths Approach to Noise-Induced Synchronization: Stochastic Resonance in a Double-Well Potential." *Annals of Applied Probability* 12(4): 1419–1470.

Berglund, N., and B. Gentz. 2002b. "Pathwise Description of Dynamic Pitchfork Bifurcations with Additive Noise." *Probability Theory and Related Fields* 122:341–388.

Bhattacharya, R., and M. Majumdar. 1980. "On Global Stability of Some Stochastic Economic Processes: A Synthesis," in L. R. Klein, M. Nerlove, and S. C. Tsiang, eds., *Quantitative Economics and Development.* New York: Academic.

Brock, W. 2004. "Tipping Points, Abrupt Opinion Changes, and Punctuated Policy Change." http://www.ssc.wisc.edu/~wbrock.

Brock, W. 2006. "Tipping Points, Abrupt Opinion Changes, and Punctuated Policy Change," pp. 47–77. In R. Repetto, ed., *Punctuated Equilibrium and the Dynamics of U.S. Environmental Policy.* New Haven: Yale University Press.

Brock, W., and S. Durlauf. 2001a. "Discrete Choice with Social Interactions." *Review Of Economic Studies* 68:235–260.

Brock, W., and S. Durlauf. 2001b. "Interactions-Based Models," in J. Heckman and E. Leamer, eds., *Handbook of Econometrics V*, pp. 3297–3380. Amsterdam: North-Holland.

Brock, W., S. Durlauf, and K. West. 2003. "Policy Evaluation in Uncertain Economic Environments." *Brookings Papers on Economic Activity* 1(2003): 235–301.

Brock, W., S. Durlauf, and K. West. 2004. "Model Uncertainty and Policy Evaluation: Some Theory and Empirics." http://www.ssc.wisc.edu/~wbrock.

Brock, W., S. Durlauf, and K. West. 2007. "Model Uncertainty and Policy Evaluation: Some Theory and Empirics." *Journal of Econometrics* 136(2): 629–664.

Broecker, W. 1987. "Unpleasant Surprises in the Greenhouse?" *Nature* 328:123–126.

Campbell, J., A. Lo, and A. MacKinlay. 1997. *The Econometrics of Financial Markets*. Princeton: Princeton University Press.

Carpenter, S. R. 2003. *Regime Shifts in Lake Ecosystems: Pattern and Variation*. Oldendorf: International Ecology Institute.

Carpenter, S.R., and W.A. Brock. 2004. "Spatial Complexity, Resilience, and Policy Diversity: Fishing on Lake-Rich Landscapes." *Ecology and Society* 9(1): 8. http://www.ecologyandsociety.org/vol9/iss1/art8.

Carpenter, S. R., and J. F. Kitchell, eds. 2003. *The Trophic Cascade in Lakes*. Cambridge: Cambridge University Press.

Carpenter, S. R., D. Ludwig, and W. Brock. 1999. "Management of Eutrophication for Lakes Subject to Potentially Irreversible Change." *Ecological Applications* 9(3): 751–771.

Conley, T. G., L. P. Hansen, E. G. J. Lutlmer, and J. A. Scheinkman. "Short-term Interest Rates as Subordinated Diffusions." *Review of Financial Studies* 10(3): 525–577.

Crepin, A., and J. Norberg. 2003. "The Paradox of Biodiversity." Beijer Working Paper. Beijer Institute of Ecological Economics, and Department of Systems Ecology, Stockholm, Sweden. http://www.beijer.kva.se.

Dawson, D. 1983. "Critical Dynamics and Fluctuations for a Mean-Field Model of Cooperative Behavior." *Journal of Statistical Physics* 31(1): 29–85.

Foster, D., and D. Nelson. 1996. "Continuous Record Asymptotics for Rolling Sample Variance Estimators." *Econometrica* 64:139–174.

Hansen, L. 1982. "Large Sample Properties of Generalized Method of Moments." *Econometrica* 50:1029–1054.

Hansen, L., and J. Scheinkman. 1995. "Back to the Future: Generating Moment Implications for Continuous-Time Markov Processes." *Econometrica* 50: 1269–1288.

Hommes, C. 2005. *Heterogeneous Agents Models: Two Simple Examples*. Amsterdam: CeNDEF, University of Amsterdam. http://www1.fee.uva.nl/cendef/publications/papers/Udine2005.pdf.

Horsthemke, W., and R. Lefever. 1984. *Noise-Induced Transitions: Theory and Applications in Physics, Chemistry, and Biology*. Berlin: Springer.

Jeppesen, E., J.P. Jensen, P. Kristensen, M. Søndergaard, E. Mortensen, O. Sortkjaer, and K. Olrik. 1990. "Fish Manipulation as a Lake Restoration Tool in Shallow, Eutrophic, Temperate Lakes 2: Threshold Levels, Long-Term Stability and Conclusions." *Hydrobiologia* 200/201:219–228

Kirkpatrick, S., C. Gelatt, and M. Vecchi. 1983. "Optimization by Simulated Annealing." *Science* 220:671–680.

Kleinen, T., H. Held, and G. Petschel-Held. 2003. *The Potential Role of Spectral Properties in Detecting Thresholds in the Earth System: Application to the Ther-*

mohaline Circulation. Potsdam: Potsdam Institute for Climate Impact Research (PIK).

Kuznetsov, Y. 1995. *Elements of Applied Bifurcation Theory*. Berlin: Springer.

Lindzen, R. 1992. "Global Warming: The Origin and Nature of the Alleged Scientific Consensus." *Cato Review of Business and Government* vol. 15, no. 2. http://www.cato.org/pubs/regulation/regv15n2/reg15n2g.html.

Ludwig, D., S. R. Carpenter, and W. Brock. 2003. "Optimal Phosphorous Loading for a Potentially Eutrophic Lake." *Ecological Applications* 13:1135–1152.

Madigan, D., and A. Raftery. 1994. "Model Selection and Accounting for Model Uncertainty in Graphical Models Using Occam's Window." *Journal of the American Statistical Association* 89:1535–46.

Magee, S., W. Brock, and L. Young. 1989. *Black Hole Tariffs and Endogenous Policy Theory*. Cambridge: Cambridge University Press.

Magill, M. 1977. "A Local Analysis of N-sector Capital Accumulation Under Uncertainty." *Journal of Economic Theory* 15:211–218.

Malliaris, A. 1982. *Stochastic Methods in Economics and Finance*. Amsterdam: North Holland.

Meijer, M. L., E. Jeppesen, E. Van Donk, B. Moss, M. Scheffer, E. H. R. R. Lammens, E. H. Van Nes, J. A. Berkum, G. J. De Jong, B. A. Faafeng, and J. P. Jensen. 1994. "Long-Term Responses to Fish-Stock Reduction in Small Shallow Lakes: Interpretation of Five-Year Results of Four Biomanipulation Cases in the Netherlands and Denmark." *Hydrobiologia* 276:457–466.

Meyer, S. L. 1975. *Data Analysis for Scientists and Engineers*. New York: Wiley.

National Research Council. 2002. *Abrupt Climate Change: Inevitable Surprises*. Washington, DC: National Academy Press.

Norberg, J., D. Swaney, J. Dushoff, J. Lin, R. Casagrindi, and S. Levin. 2001. "Phenotypic Diversity and Ecosystem Functioning in Changing Environments: A Theoretical Framework." *Proceedings of the National Academy of Sciences* 98(20): 11376–11381.

Sarewitz, D., R. A. Pielke Jr., and R. Byerly Jr. 2000. *Prediction: Science, Decision Making, and the Future of Nature*. Washington, DC: Island.

Scheffer, M. 1998. *The Ecology of Shallow Lakes*. London: Chapman and Hall.

Scheffer, M., S. R. Carpenter, J. A. Foley, C. Folke, and B. Walker. 2001. "Catastrophic Shifts in Ecosystems." *Nature* 413:591–596.

Scheffer, M., F. Westley, and W. Brock. 2000. "Socioeconomic Mechanisms Preventing Optimum Use of Ecosystem Services: An Interdisciplinary Theoretical Analysis." *Ecosystems* 3:451–471.

Scheffer, M., F. Westley, and W. Brock. 2003. "Slow Response of Societies to New Problems: Causes and Costs." *Ecosystems* 6:493–502.

Steele, J. H. 1998. "Regime Shifts in Marine Ecosystems." *Ecological Applications* 8:S33-S36.

Stow, C. S., S. S. Qian, and J. K. Craig. 2005. "Declining Threshold for Hypoxia in the Gulf of Mexico." *Environmental Science and Technology* 39:716–723.

Van Geest, G. J., F. Roozen, H. Coops, R. M. M. Roijackers, A. D. Buijse, E. Peeters, and M. Scheffer. 2003. "Vegetation Abundance in Lowland Flood Plan Lakes Determined by Surface Area, Age and Connectivity." *Freshwater Biology* 48:440–454.

Wood, B., and A. Doan. 2003. "The Politics of Problem Definition: Applying and Testing Threshold Models." *American Journal of Political Science* 47(4): 640–653.

Woodwell, G., and F. MacKenzie, eds. 1995. *Biotic Feedbacks in the Global Climate System: Will the Warming Feed the Warming?* Oxford: Oxford University Press.

INTRODUCTION TO PART 4
PRACTICAL APPROACHES

SO FAR, THIS BOOK has taken you on a tour of many of the theories and generalities that underly the study of complex adaptive systems. Although the preceding chapters touch on a wide range of important applications of these ideas in the arenas of management and policy, most of them do not address applications directly. This final section is intended to demonstrate some of the practical approaches that have been developed to facilitate our interactions with complex adaptive systems.

One of the questions of greatest interest to many of the authors who have contributed to this volume is that of how we as humans can live sustainably in a world that contains finite resources. Many of the environmental problems that we are now starting to see direct evidence of—such as climate change, pollution, and habitat destruction—are consequences of mass action. That is to say, they are problems not so much because each individual action is harmful, but because a sufficiently large number of people are performing actions like driving cars, cutting trees, and dumping their unprocessed sewage into water bodies. Such excess overwhelms the capacity of ecosystems to process pollutants, regulate the climate, or purify water. If we are to effectively work toward a future in which both humans and the ecosystems we depend on are maintained as parts of the global system, we need to find ways of influencing the actions of large numbers of people to manage the environment in a way that is responsive, rather than under a set of locally optimal rules that quickly become inappropriate when the environment changes, and to understand which actions will have broad-scale consequences.

The first chapter in this section deals with people and their perceptions of the future. It presents a set of methods that can help societies to look forward and develop appropriate management and policy responses to environmental change, based not only on current trends but also on careful consideration of the ways in which the future may differ from the past and the present. Stakeholders, the people who have some kind of ownership within a social-ecological system, obtain benefits from it, but have the additional responsibility to manage it appropriately. We are all global stakeholders. When scientists work with local stakeholders to help develop appropriate policy and management solutions, the success of their efforts is contingent not only on the effectiveness of the solution, or the quality of the science that underlies it, but also on a set of political variables that include the current leadership, governance institutions, and the willingness (and ability) of the local community to learn from science and apply the best solutions. Consequently, any attempt to work toward sustainability must consider the involvement of stakeholders at different scales and must pay careful attention to social and political processes.

One of the frameworks that at present is most in vogue as a way of achieving successful management in a changing environment is adaptive management (or adaptive comanagement, when it involves partnerships between multiple organizations). The second chapter in this section takes a long, hard look at adaptive management in the context of complex adaptive systems theory. The authors explain what adaptive management is, ask how successful it is, and discuss some of the lessons that we have learned from adaptive management attempts.

The third and final chapter of this section brings us back to one of the central problems in working with complex adaptive systems; that of scale. It reviews the scaling literature and then summarizes some of the many scale-related issues that have arisen through the book. How can these and related ideas about scale and scaling be useful in managing complex adaptive systems? Monitoring is an essential part of management, and we learn about the behavior of real-world systems by observation. Adaptive management and scenario planning in their strongest forms depend heavily on modeling, which must inevitably confront scale directly. Coming to grips with scale is therefore essential for both measurement and modeling of complex adaptive systems. Scale also becomes an important consideration for managers and policy makers when considering interventions in complex systems, such as economies or ecosystems. If the institutions that are responsible for managing a process are incapable of manipulating that process at the necessary scale, then management will be ineffective. As this chapter makes clear, coping with scale-related issues in complex adaptive systems is a major challenge from both theoretical and applied perspectives.

7

PARTICIPATION IN BUILDING SCENARIOS OF REGIONAL DEVELOPMENT

Louis Lebel and Elena Bennett

THE PAST IS NOT AN infallible guide to the future, but is shaped rather by the way people think about it from their present. Looking forward, the human activities and landscapes that define a region could conceivably unfold in several different ways. Scenarios have long been a tool of business and military strategists for exploring situations of uncertainty. Scenarios are stories about how the future may unfold. They can be qualitative and quantitative, told looking forward from today or looking backward from an envisioned future, and can be used to examine systems over long time frames (van Notten et al. 2003). Scenario-building approaches vary with purpose and have shifted from planning focused on assessing vulnerabilities or threats to a process of learning and building that also looks for opportunities (Masini and Vasquez 2000; Neumann and Overland 2004). They are increasingly used in studies of environmental change and development (Gallopin et al. 1997; Swart, Raskin, and Robinson 2004).

Building scenarios usually involves expert input and desk research (Neumann and Overland 2004). Some proponents also seek broader stakeholder inputs or public participation. Participation is interpreted in many ways (International Association for Public Participation 2000). It can vary from being little more than telling others about the findings to active involvement in setting goals, refining stories, and drawing policy recommendations. In this chapter we explore the value and challenges of expanding participation in building scenarios of regions. The chapter is organized into four sections. The first looks at why participation is sought. The second looks at who participates in which roles. The third explores how participants are engaged. And the fourth looks at the content of the resulting scenario "products."

RATIONALES

Conveners are those who coordinate, facilitate, and in other ways support a scenario-building exercise. Why do conveners recruit other participants to scenario-building exercises? Scenarios of regional futures are made for different purposes. As a consequence their rationales for different levels and forms of participation differ (table 7.1). We contrast three.

First, if the main goal is to get an alternative strategy or policy option for regional development into the public domain, then conveners will focus on recruiting supporters and influential players that need to be turned into supporters. This rationale is rarely made explicit because it is so obviously strategic. Nevertheless, we believe it is a common approach.

Second, if the main goal is to better understand the underlying drivers, how actors might respond, and key uncertainties then conveners will mostly focus on getting a good cross-section of knowledgeable individuals that includes technical experts. A complex system perspective would encourage expanding inputs beyond academics to target groups and policy makers or bureaucratic managers who can bring in their practical knowledge of the system.

Third, if the main goal is to facilitate social learning across stakeholder groups then conveners will try to ensure diverse interests are represented. Including people who could be affected by decisions arising out of deliberations is a priority, but much wider involvement may be encouraged. Experts may also be called upon, but here they would primarily be consultants to the process rather than leaders. Practical constraints of cost and time can be telling for depth of public participation actually achieved.

There is a place for all three approaches to participation in building scenarios. The claims made by conveners regarding participation, however, de-

TABLE 7.1 Rationales of Conveners for Participation in Building Scenarios

RATIONALES OF CONVENERS	KEY PARTICIPANTS SOUGHT	PUBLIC PARTICIPATION MODE
Introduce alternatives	Supporters—articulate advocates Door openers	Inform
Improve understanding	Experts, target groups	Consult
Facilitate social learning	Stakeholders, consulting experts	Empower

serve scrutiny, especially if they are intended to imply "representativeness" of findings. This is particularly important when the convener is a "state" agency with relatively direct links to decision-making processes. Promises are not always kept.

Finally we must underline that conveners and invited participants may not share identical views regarding participation. Thus while conveners often have instrumental reasons to expand participation, at least as far as costs allow, stakeholders may wish to be included so that their interests are properly represented or, alternatively, to have competitors excluded. On the other hand, some candidate participants may opt not to become involved because they feel an exercise is a waste of time or because they do not want to be seen as legitimizing a process that is token or otherwise seriously flawed.

PLAYERS

Building scenarios potentially involves actors playing several roles—at the most basic level, conveners and participants.

Conveners coordinate, facilitate, and support scenario-building exercises. They are typically the ones holding a project grant, although they may employ independent consultants for certain tasks. The motivation, interests, and skills of conveners may matter more for participation than the final composition of participant lists.

As facilitators conveners need to be ethical, honest, and respectful of views of others. The dynamics of conversations can also lead to participants identifying strongly with individual scenarios in ways that prevent rather than enhance opportunities for learning about the consequences of various assumptions behind them. For this reason skills in facilitation are important in order to encourage a constructive debate, an understanding of others, and to avoid domination by individuals or small groups.

Facilitators can work to empower weaker parties and ensure that everyone can play a role in the scenario development process. As with all multistakeholder activities, a challenge for conveners is handling power relations among stakeholders. This may require dividing groups and providing special assistance to disadvantaged groups (Wollenberg, Edmunds, and Buck 2000a). In scenario development, facilitators continuously challenge plausibility, ask about consequences and preconditions, and encourage articulation of assumptions. When the goal is an informative set from which contrasts will be drawn, facilitators often need to help amplify contrasts between scenarios.

Participants are those who accept invitations to attend the workshops, meeting, or public Web pages where they can submit their inputs. They are potentially a very diverse group. A few are almost always critical for understanding regional social-ecological systems.

Resource users (e.g., fishers, farmers) are an important category. They often include individuals who are keen observers of trends and relationships, in part because they have to be to make their livings. Their inclusion in building scenarios even when they are not the primary target group can be important for understanding of potential thresholds, feedbacks, and other sources of nonlinearity

Resource managers are a second important group, often in conflict with the first. As members of agencies with authority to make decisions or carry out operations that impact others, their involvement is crucial; they often hold the "scientific databases" and can help test plausibility of monitoring or alternative management.

Researchers have played a key critical role in the early detection and basic understanding of problems of unsustainability. Scientists can bring knowledge about processes, relations, thresholds, or other important factors that are not immediately obvious but important to bear into scenarios (Swart, Raskin, and Robinson 2004).

The wider public, without specialist knowledge or immediate commercial "stake," can also contribute, especially in helping link narrower concerns that drive the creation of scenario exercise in the first place with other problems and opportunities in society. People can feel responsibility for environmental sustainability and social justice problems and want to be part of identifying solutions.

Finally, there are the sponsors that commission, pay for, or otherwise support the building of scenarios. They may be the same or closely related to the conveners or intentionally remain distant from the process until it is completed. Whether it is a state or nongovernmental organization can make a lot of difference in how the public and polity treat an exercise. Their continued support or withdrawal can be critical to the impacts an activity has either way.

BUILDING

Building scenarios together requires attention to issues of when and how participants are engaged.

Conveners seek participation at different stages. Early on participation can help define boundaries in time and space, the main issues and outcome variables that have to be understood, and the key uncertainties that will be considered. Early participation can help refine the purpose of the overall exercise and create the enthusiasm and commitment needed for a successful iteration. More typically, conveners make most of the early scoping decisions themselves, for example, as part of proposals for grants, and then market and defend these decisions. This practice can undermine perceived legitimacy. Participation is then sought in the middle stages where causal relationships are teased out and alternative trajectories explored. This may involve iterations among a smaller group that rewrites (or rebuilds) the scenarios and a wider group that validates and challenges. Finally, participation can be sought in the final stages where interpretations of scenarios are refined, response options explored, and the implications for current development strategies debated.

At each of these stages many different forms of public participation are conceivable and several have been tried.

Wollenberg, Edmunds, and Buck (2000a, 2000b) focus on using scenarios to enable stakeholders to learn and to act in new ways. To meet this goal, they deem that matching participants, facilitator, and mode of communication to the purpose of the scenarios is critical. The most appropriate media for conducting scenario exercises depends greatly on text. Computer screens and maps may not be at all appropriate in a village setting, but poems or stage plays might be. As stakeholders often vary greatly in their communication skills and power relations shift with venue, a lot can be said for using a diversity of approaches and avenues of input in scenario building.

The key appears to be creating an atmosphere conducive to open dialogue where people are able to express dissent and contrary views with safety. People need to feel they can play a role in both challenging the internal consistency and plausibility of a scenario and, more fundamentally, in determining whether the contrasts implied are even useful. They are most likely to engage if they feel that their views are listened to and they, in turn, are learning from what others have to say. Learning occurs both by being involved in the construction of the scenario and, just as important, by discussing and exchanging views with others about the elements and implications of the scenarios. Subject matter experts may be called in as advisers without being directly involved in the scenario building. Crucially, there is a sense that knowledge arising is coproduced. Scenarios with participation are social activities and as such have potential for fostering social learning and building trust. Learning is fundamentally iterative. Learning loops through which the group cycles via discussions of uncertainties, articulating assumptions, assessing consequences, and reflecting on both findings and process may be particularly effective.

On the other hand, building scenarios together in a diverse group may lead to gridlocks and polarization of views aligned behind alternative futures rather than better or shared understandings of assumptions. Wollenberg, Edmunds, and Buck (2000a) summarize a case study from southern Zimbabwe where an influential ex-local government leader trying to derail the process had to be taken to an alternative meeting to discuss important business to allow others a chance to have a say. Conveners may intentionally exclude or separate participants because of costs, disruptive influence, or to protect their own agendas.

A rather different approach was used in a transport study. In a study of urban transport futures in the Netherlands responses of experts to a common questionnaire were used to get composite views of how eight major classes of driving factors were expected to change (Nijkamp, Ouwersloot, and Rienstra 1997). A second scenario was derived from questions about how drivers should change: a normative scenario. The narrow group of participants was very uniform in their responses. Emissions of carbon dioxide were calculated under the two scenarios illustrating that expectations were essential of no reductions and that "what was wished" was a long way away. This is a rather extreme example of standardizing inputs of participants, but for an admittedly narrow group of transport experts.

The Georgia Basin Futures Project drew on expert knowledge and community inputs to build tools and a game for exploring what-if type scenarios for a basin on the west coast of Canada (Tansey et al. 2002). A backcasting approach is used in which participants pose futures and then trade-offs in achieving it are explored (Robinson 2003).

Scenario building has emerged as a loose collection of approaches without much theoretical underpinning or critical evaluation of methods (Chermack 2005; Neumann and Overland 2004). The diversity of approaches is staggering and promising but needs systematic analysis that can be the basis of improvement of methods. Unfortunately, most scenario-building activities are poorly documented. A notable exception is a participatory study of biomedicine in Germany that carefully documents the choice of participants, the process for deriving the scenarios, and a final evaluation by participants (Niewohner et al. 2005). Another illustration is the MedAction project, which used the VISIONS scenarios as a starting framework and involved a similar diversity of participants in three stakeholder workshops (de Groot et al. 2004). In selecting participants, consideration was given to many factors including age and background, but ultimately depended a lot on the local organizers in the target areas. At the first workshop farmers did not turn up, and this was corrected in later meetings. An extreme case of dominance by a small group was documented by the MedAction group in one of their scenario modules, but because care was taken to record the process this could be taken into account when interpreting findings from different stakeholder workshops later.

CONTENT

So far we have largely concentrated on process. The content of a scenario-building exercise, as embodied in the ideas discussed, knowledge brought to bear, and visions shared is also shaped by participation. This is most visible in the final products (books, films, drawings) of a scenario-building exercise, but is not confined to these. Participation appears useful for understanding causal clusters, identifying nonlinearities, handling cross-scale interactions, and exploring adaptive capacities—key components of scenarios of regional development in an uncertain world.

Scenario can be useful for building an understanding of a complex system because they force participants to confront assumptions and knowledge about how the system works. For example, when the causal structure of a system is difficult to unravel, scenarios can help identify clusters of causal variables that operate together.

A good example comes from the VISIONS project, a well-resourced, three-year scenario exercise carried out on behalf of the European Commission (van Notten and Rotmas 2001). It developed qualitative story lines for alternative futures of Europe, which were supported by quantitative information and simulation modeling. Integrating the two kinds of information was difficult (van Notten et al. 2003). More focused scenarios were also initially prepared for three smaller regions with Europe. The scenarios were relatively complex and involved pathways of change for twelve interacting variables and feedback mechanisms. Story lines for both Europe and smaller regions were developed through multistakeholder workshops with diverse participants including representatives from the science community, businesses, government institutions, NGOs, citizens, and artists from various EU member states (van Notten et al. 2003). The Directorate General of the EU, which commissioned the scenario exercises, however, remained only passively involved (van Notten et al. 2003). This contrasts with the experience of the same group in the Netherlands in a 2030 scenario project commissioned by the Dutch minister of housing, spatial planning, and environment. After presenting three descriptive and exploratory scenarios to the minister, the group was requested to add a fourth normative scenario that included current strategy, effectively constraining the initial exploratory intent of the exercise (van Notten et al. 2003).

Scenario planning is also useful when the pathway to the future is nonlinear or when the potential for novelty or surprise is high. A long-term partnership between the Goulburn Broken Catchment Authority and various resource users, managers, and experts from a government-applied research agency led to an initial technical study inventorying ecosystem services and was followed by the policy-oriented assessment (Binning et al. 2001; CSIRO Sustainable Ecosystems 2003). Scenarios focused on different vegetation management strategies were used to assess changes in ecosystems services from a dryland subcatchment. The presence of sufficient ecological and spatial detail of the supporting models allowed nonlinearity in the provision of services as a result of neighborhood effects to be explored. Trade-offs could also be identified by comparing scenarios.

Novelty or innovation can be both positive and negative from the perspective of sustainability and can take several forms in scenarios. The introduction of a novel technology that greatly enhances capacities of societies to harvest or exploit a resource or substitute such exploitation with a renewable alternative are of particular importance. Over long times these might be captured in ideas of rates of innovation reflecting investments in science and technology for sustainability. Less frequently considered but also worth exploring are institutional novelties, for example, with respect to global governance or monitoring of environment and what consequences this might have. Kahane (1998) introduced the ideas of novelty through "wildcards" in scenario discussions about future of South Africa. Surprises by definition cannot be directly planned for. An unexpected disease outbreak affecting recreational fisheries is included in one of the scenario story lines for a lake-dominated landscape in Wisconsin (Peterson et al. 2003). Surprise or novelty may act as triggers of change in some story lines, whereas in others they prevent a particular course from being pursued to its completion. Diverse participation should increase the chances that significant but plausible novelties arise to be considered.

Most regional scenario exercises must deal with issues of scale, but mostly do so by collecting these into assumptions about external conditions. The Millennium Ecosystem Assessment was explicitly a multiscale assessment aimed at capturing social-ecological linkages in detail. It provided opportunities for subglobal regional assessments to contrast their own local scenarios with the overarching set of global scenarios (Millennium Ecosystem Assessment 2005).

Complex systems approaches suggest that for interesting behavior to emerge at a regional level of interest it is not essential that all the details of local interactions be tracked. Scenarios may emphasize patterns at the higher level without needing to pose unrealistic levels of omniscient control of events, rather explaining these changes based on behavior of smaller interacting elements. For example, regional stakeholders interact with other stakeholders through networks, contesting resources, cooperating in their management, and sharing understanding about past and potential future ecological and social changes. These interactions among actors, and changes in delivery of goods and services from ecosystems they are using, may feed back to influence those behaviors in a self-organizing way (Lansing 2003; Levin 2003). A consequence may be that emergent patterns arise at the regional level even in the absence of highly centralized coordination or control (Berkes, Colding, and Folke 2003; Holland 1995).

Scenarios encourage people to prepare for the future as something different than the immediate past. This can increase attention given to building and maintaining adaptive capacities that otherwise would be forgotten (Wollenberg, Edmunds, and Buck 2000b). Once developed, scenarios may also be used to explore the robustness of key management decisions or development strategies. This is accomplished by examining how key management decisions would play out in each scenario and is often done to make sure that management decisions will be robust no matter what the future holds. Building scenarios together may help trigger self-organization by in-creasing awareness and preparedness for contingencies.

The form and extent of public participation in building scenarios of re-gional development matters for what issues get addressed, which sources of knowledge are drawn on, and use and impact.

Decorative participation has no purpose other than to give the appear-ance of having consulted with a wider group. As such it is deceitful. Some-times there are good reasons not to consult or engage with a wider group of stakeholders in a scenario-building exercise. Conveners should not be frightened of being honest. *Meaningful participation*, on the other hand, has as its goal empowering those involved to alter the dynamics, content and impact of building scenarios. It can do so in several ways, but each brings with it important challenges.

First, having diverse inputs means more factors, interactions, and inter-ests. It also means that less well-known thresholds, feedback, and trends can be tabled, though these may be ignored. Bringing them together into internally consistent and plausible individual scenarios, however, is likely to be much harder than with fewer inputs. The temptation is dilution. Ne-gotiating a set of scenarios that will yield informative contrasts gets harder as well.

Second, social interactions can build understanding and trust without necessarily leading to consensus. It changes the way people think about the future, thus, conceivably, reshaping the future. People who are always chal-lenging and thinking about uncertainties demand more attention be paid to adaptive capacities rather than one-time-only solutions. This can be very difficult for bureaucracies, built on conventional division of labor, the fear of short-term election cycles, and fixed mandates, to bring more adaptive approaches into their management strategies and operations. It can also be difficult for citizens. But ultimately this shift to embracing uncertain-ties and thinking more about adaptive capacities is essential for regional sustainability.

Third, compared to a firm with a relatively narrow set of goals, regional social-ecological systems are extremely heterogeneous. There are more goals and interests but fewer levers for management to pull. Building scenarios in this latter context is inherently a different activity than it is in a firm or military strategy group. The outcomes are more ambiguous and plausible response options broader. The links with decision making are usually indirect and vague. In this situation building scenarios is one tool (among several) for supporting "dialogues" rather than a more narrowly construed "decision-support" tool, as it has often been framed in other fields. But the tension between exploring and giving advice is real and belongs in the realm of politics.

With meaningful participation scenario-building exercises become much more than a variant of conventional approaches to strategic planning because they turn the activity into a platform for dialogue. Scenario-building is a plausible way for stakeholders, researchers, and decision makers to engage with the complexity of regional development and explore alternative futures.

REFERENCES

Louis Lebel thanks the Rockefeller Foundation and the U.S. National Science Foundation for their support. Elena Bennett thanks the Center for Limnology, University of Wisconsin, and the Millennium Ecosystem Assessment, as well as the many participants in scenario development exercises she's been involved in.

Berkes, F., J. Colding, and C. Folke, eds. 2003. *Navigating Social-Ecological Systems: Building Resilience for Complexity and Change.* Cambridge: Cambridge University Press.

Binning, C., S. Cork, R. Parry, and D. Shelton. 2001. *Natural Assets: An Inventory of Ecosystem Goods and Services in the Goulburn-Broken Catchment.* Canberra: CSIRO Sustainable Ecosystems.

Chermack, T. J. 2003. "Mental Models in Decision Making and Implications for Human Resource Development." *Advances in Developing Human Resources* 5:408–422.

Chermack, T. J. 2005. "Studying Scenario Planning: Theory, Research Suggestions and Hypotheses." *Technological Forecasting and Social Change* 72:59–73.

CSIRO Sustainable Ecosystems. 2003. *Natural Values: Exploring Options for Enhancing Ecosystem Services in the Goulburn Broken Catchment.* Canberra: CSIRO Sustainable Ecosystems.

de Groot, D., K. Kok, M. Patel, and D. S. Rothman. 2004. *Manual on Policy Analysis for the Mitigation of Desertification.* MedAction Deliverable 37. ICIS Working Paper: IO4-E003. Maastricht: ICIS, Maastricht University.

Gallopin, G., A. Hammond, P. Raskin, and R. Swart. 1997. *Branch Points: Global Scenarios and Human Choice.* Stockholm: Global Scenarios Group, Stockholm Environment Institute.

Holland, J. H. 1995. *Hidden Order: How Adaptation Builds Complexity.* New York: Helix.

International Association for Public Participation. 2000. *Public Participation Spectrum*. International Association for Public Participation. http://www.iap2.org/associations/4748/files/spectrum.pdf.

Kahane, A. 1998. "Imagining South Africa's Future: How Scenarios Helped Discover Common Ground." In L. Fahey and R. M. Randall, eds., *In Learning from the Future: Competitive Foresight Scenarios*, pp. 325–332. New York: Wiley.

Lansing, J. S. 2003. "Complex Adaptive Systems." *Annual Review of Anthropology* 32:183–204.

Levin, S. 2003. "Complex Adaptive Systems: Exploring the Known, the Unknown and the Unknowable." *Bulletin of the American Mathematics Society* 40:3–19.

Masini, E. B., and J. M. Vasquez. 2000. "Scenarios as Seen from a Human and Social Perspective." *Technological Forecasting and Social Change* 65:49–66.

Millennium Ecosystem Assessment. 2005. *Ecosystems and Human Well-being: Synthesis*. Washington, DC: Island.

Neumann, I. B., and E. F. Overland. 2004. "International Relations and Policy Planning: The Method of Perspectivist Scenario Building." *International Studies Perspectives* 5:258–277.

Niewohner, J., P. Wiedemann, C. Karger, S. Schicktanz, and C. Tannert. 2005. "Participatory Prognostics in Germany: Developing Citizen Scenarios for the Relationship Between Biomedicine and the Economy in 2014." *Technological Forecasting and Social Change* 72:195–211.

Nijkamp, P., H. Ouwersloot, and S. A. Rienstra. 1997. "Sustainable Urban Transport Systems: and Expert-Based Strategic Scenario Approach." *Urban Studies* 34:693–712.

Peterson, G. D., T. D. J. Beard, B. Beisner, E. Bennett, S. R. Carpenter, G. S. Cumming, C. L. Dent, and H. T. D. 2003. "Assessing Future Ecosystem Services: A Case Study of the Northern Highland Lake District, Wisconsin." *Conservation Ecology* 7:1. http://www.ecologyandsociety.org/vol7/iss3/art1/.

Robinson, J. 2003. "Future S: Backcasting as Social Learning." *Futures* 35:839–856.

Swart, R., P. Raskin, and J. Robinson. 2004. "The Problem of the Future: Sustainability Science and Scenario Analysis." *Global Environmental Change* 14:137–146.

Tansey, J., J. V. Carmichael, R. Van Wynsberghe, and J. Robinson. 2002. "The Future Is Not What It Used to Be: Participatory Integrated Assessment in the Georgia Basin." *Global Environmental Change* 12:97–104.

van Notten, P., and J. Rotmas. 2001. "The Future of Scenarios." *Scenario and Strategy Planning* 1:4–8.

van Notten, P. W. F., J. Rotmans, M. B. A. van Asselt, and D. S. Rothman. 2003. "An Updated Scenario Typology." *Futures* 35:423–443.

Wollenberg, E., D. Edmunds, and L. Buck. 2000a. "Anticipating Change: Scenarios as a Tool for Adaptive Forest Management." Bogor: Center for International Forestry Research (CIFOR). http://www.cifor.cgiar.org/.

Wollenberg, E., D. Edmunds, and L. Buck. 2000b. "Using Scenarios to Make Decisions About the Future: Anticipatory Learning for the Adaptive Co-management of Community Forests." *Landscape and Urban Planning* 47:65–77.

8

PRACTICING ADAPTIVE MANAGEMENT IN COMPLEX SOCIAL-ECOLOGICAL SYSTEMS

Lance Gunderson, Garry Peterson, and C. S. Holling

FOR THOUSANDS OF years humans have purposefully and inadvertently modified ecosystems through the manipulation of ecological processes and ecosystem structures. Agriculture, for example, attempts to alter ecosystem goods such as biodiversity and ecosystem services such as water movement in order to increase specific types of ecological production. Yet, as humans increase the supply of desired ecosystem services by simplifying and homogenizing ecosystems, unwanted side effects emerge. The simplification and stabilization of these systems tends to reduce their ability to reliably supply services, even as human society becomes more dependent upon their presence.

This pattern of ecosystem modification has been described as the pathology of resource management—where simplifying and stabilizing ecosystems has the unintended consequence of increasing their vulnerability (Holling and Meffe 1996). The pathology results from the unexpected response of complex systems to simple management approaches and has been increasingly problematic as both the scope of humanity's alteration of the biosphere (Vitousek 1997) and the number of people dependent upon the reliable supply of ecosystem services has increased (Steffen et al. 2004). Coping with the complex ecological changes humanity intentionally and unintentionally makes has now become a major focus and challenge for environmental management.

Complex resource systems behave in surprising ways (Holling 1978), one of which has been described as flips among alternate states or regimes (Holling 1973). Regime shifts induced by human interventions cross the spectrum of ecosystems from marine to freshwater to terrestrial (Gunderson and Pritchard 2002; Scheffer et al. 2001; Folke et al. 2004). Documented regime shifts include the transition from grass to shrub dominance in semi-arid rangelands (Walker et al. 1981), population outbreaks of forest pests (Ludwig, Jones, and Holling 1978; Holling 1986), shifts between clear water and turbid shallow lakes (Carpenter, Ludwig, and Brock 1999; Carpenter, Brock, and Hanson 1999b; Scheffer 1998), and from coral to algae dominated reefs (Hughes et al. 2004) among others. In all of these cases the transition or nonlinearity is mediated by the interaction between slower and faster components in ecosystems (Holling and Gunderson 2002).

The dynamics of complex resource systems are difficult to predict (Holling 1978; Walters 1986), much less manage (Gunderson 2003). The ecological regime shifts presented in the previous paragraph are illustrative of one source of that unpredictability. Environmental planning and management, however, require some estimation about "what will happen." Certainly, many things are known, especially the broad and the general. To cite a recent example, it was well known at least three days prior to landfall that Hurricane Katrina was going to strike the Gulf Coast of the U.S. (with a given probability), yet all the impacts could not be specifically predetermined. There are many reasons why our predictive abilities of ecological systems are limited. These include the evolutionary and cross-scale nature of ecosystems, scientific disagreement about how ecological systems are structured and function, and lack of data/information to test ideas about ecosystems across scales. Furthermore, human action has so transformed the biosphere that it is unclear how well the ecological past can serve as a model of the future (Steffen et al. 2004).

Environmental management must confront the complexity of social-ecological systems. One source of complexity is the dynamic nature of the ecological, economic, social, political, and organizational components of these systems (Holling and Gunderson 2002; Westley 2002) as well as the interaction among these components. The difficulties in managing the interacting aspects of social-ecological systems have led some to term them wicked problems (Rittel and Webber 1973; Ludwig, Mangel, and Haddad 2001). Developers of adaptive management (Holling 1978; Walters 1986) acknowledged the complex multidimensionality of natural resource issues, but focused on analytic approaches primarily in the ecological and economic domains from a systems perspective. Lee (1993) was the first to separate these issues into scientific (primarily ecological) and social (political) components.

Adaptive management was developed as an approach to environmental management that recognized ecosystems as complex, dynamic systems

(Holling 1978; Walters 1986). Adaptive environmental assessment and management (AEAM) was proposed as a process to allow managers to confront the complexity, uncertainty, and surprises inherent in these complex systems. Adaptive management is an ongoing process that combines assessment with management actions in order to learn about the complexities of system dynamics as well as to achieve intended social objectives (Holling 1978; Walters 1986; Lee 1993). Assessing a system requires synthesizing available data to generate a set of competing alternative explanations about particular sets of resource problems and social objectives. Management actions are designed by considering which actions are robust to uncertainties among alternative explanations and which actions will help test and winnow those uncertainties (Walters 1986). Management actions are evaluated by monitoring system indicators in a process that uses that information to promote learning. While these activities are described linearly, adaptive management is typically an iterative process that develops an ongoing dialogue about the functioning of the system and the goals of management. While the approach has been around for almost thirty years, it has proven to be "technically capable and socially challenged" (Johnson 1999). One of the main challenges of adaptive management is to develop mechanisms that enable social learning and experimentation.

This book argues that considerations of information processes, networks, and diversity are key aspects of complex systems. Adaptive management has a history of addressing each of these themes. The focus of adaptive management on learning involves a formalized process of information processing that combines synthesis, experimentation, and monitoring (Walters 1986). The adaptive management process also develops social networks that link scientists, managers, and policy makers with an extended peer community (Holling and Chambers 1973; Walters 1986). Those networks are important facilitators of social learning in ecosystem management (Berkes, Colding, and Folke 2003) and in managing a diversity of functions within a social-ecological system.

We discuss how adaptive management engages these themes in the next sections. We begin with a description of the processes involved with AEAM. We use that section to move into a discussion of information processing during AEAM processes. We follow with a section on social learning, which integrates information processing, diversity, and networks through adaptive management processes. Two iconic ecosystems, the Everglades of Florida and Grand Canyon portions of the Colorado River ecosystem, are discussed in the final section in order to provide examples of lessons learned from the applications of AEAM in managing these complex resource systems.

ADAPTIVE ENVIRONMENTAL ASSESSMENT AND MANAGEMENT

Adaptive environmental assessment and management is a term for a set of processes designed to integrate learning with management actions (Holling 1978; Walters 1986; Lee 1993). These processes focus on developing hypotheses (or explanations) around specific resource issues that include 1. how specific ecological dynamics operate and 2. how human interventions will affect the ecosystem (figure 8.1).

The development of those hypotheses is achieved in a process called adaptive environmental assessment, within which a computer model is used

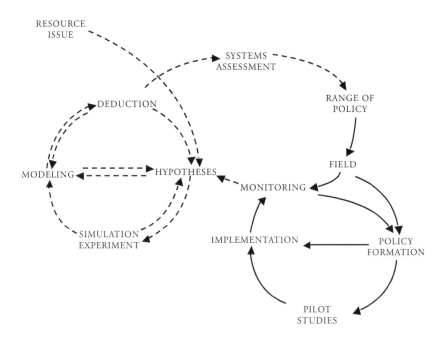

FIGURE 8.1. Conceptual model of adaptive environmental assessment and management, indicating the integration of processes that assess, propose, test, and evaluate hypotheses of ecosystem dynamics and policy implementation (modified from the original model of Holling 1981). Dashed lines represent phases of adaptive assessment, solid lines represent adaptive management.

to summarize understanding and alternative perspectives of ecosystem dynamics and evaluate a set of possible policy outcomes (Walters 1986). This process is quite different from environmental assessments routinely performed within frameworks established by governments (such as the U.S. National Environmental Policy Act of 1970) or assessments of global climate change, primarily because these approaches presume certainty of impacts and outcomes and generally assume away uncertainties. In contrast, AEAM highlights uncertainty and designs actions that will test those uncertainties about ecosystem dynamics and outcomes of proposed actions. In order to test that understanding, management actions are structured as much for learning (hence winnowing uncertainty) as they are to meet other social goals. Adaptive management views policies as hypotheses and management actions as treatments that are structured to "test" (in a weak experimental sense) desired outcomes. Walters (1986) suggested that adaptive management can be either active or passive. Active adaptive management involves direct manipulation of key ecological processes to test understanding and policies, such as the flood release experiments of 1996 in the Grand Canyon (Walters et al. 2000). Passive adaptive management uses natural variability in ecological processes to evaluate systems response, such as the experimental water delivery program in the Everglades (Light, Gunderson, and Holling 1995; Walters, Gunderson, and Holling 1992). Information gathered during the adaptive management phase is used to evaluate the hypotheses developed in the assessment phase (Walters and Holling 1990). Over time both the understanding of ecological dynamics and the impact of policies on ecosystems change (Lee 1993; Gunderson, Holling, and Light 1995; Walters 1997), leading to reiterations of the processes depicted in figure 8.1. Moving through these processes (figure 8.1) of AEAM has been described as social learning (Lee 1993), which is discussed later in the chapter. The processes of AEAM depicted in this paragraph begin with adaptive environmental assessment.

The process of adaptive environmental assessment was developed to answer three questions (Holling 1978): 1. How can complex resource issues be decomposed into a process that is tractable and understandable? 2. How can dispersed expertise and information be brought to bear on the complex resource issues? 3. How can decision makers and stakeholders in the system be engaged in learning and analysis? The first two questions relate to information processing, which involves developing a framework for simplifying the complex and synthesizing extant information. The framework and synthesis process are described in the following section on information processing. The third question of the assessment involves the broader issue of creating a social network in order to learn, which is discussed in the subsequent section on social learning.

INFORMATION PROCESSING IN AEAM

Information is the currency of the adaptive assessment and management process (figure 8.1). Information is processed through a series of workshops that are organized around constructing a computer model or models in order to help highlight what is known and not known about a resource issue. The way in which the information is processed involves a framework for integrating and synthesizing information that tests understanding of ecosystem dynamics and effects of interventions.

A number of frameworks have been developed to process information on resource issues and guide management actions. These include 1. qualitative assessments by knowledgeable persons, 2. expert judgment, 3. narratives or scenarios, 4. strategic gaming, e. statistical analyses of empirical or historical data, and f. causal modeling (National Research Council 1999). Expert opinions can be used in legal situations or where causal understanding or data or lacking. The development of narratives or scenarios has been used successfully to highlight emerging issues (Rachel Carson's *Silent Spring* is a good example) or to roughly define feasible alternative futures that contain great uncertainty. Scenario-based planning has been used in business, issues of climate change, to explore global ecological futures, and in regional resource issues (Carpenter 2002; Peterson, Cumming, and Carpenter 2003). Strategic gaming, having resource users and decision makers construct a model system in collaboration with a modeling team, was an early part of adaptive management (Holling 1978) and has been applied in a wide variety of contexts (Barreteau, Bousquet, and Attonaty 2001; Lynam et al. 2002). Statistical analysis is part of every adaptive management project as available data are analyzed or compared. Often this involves bridging scientific disciplines to gain new insight into old problems. For example, archeological data was used to assess the size and use of salmon fisheries before European contact in the Pacific Northwest (Lee 1993).

Casual modeling is the framework most frequently used in adaptive assessment and management; AEAM models provide a transparent mechanism to identify areas of understanding and uncertainty (Walters 1986). These models are intended to be minimal, credible representations used to design management actions that can test understanding of resource dynamics and potential impacts of interventions. They are not used to dictate policy choices or optimize outcomes, in contrast to other applications of causal modeling.

Other approaches to ecological management often cope with uncertainty in prediction by fitting models to data and comparing competing models (Clark et al. 2001; Hilborn and Mangel 1997). Their aim is to discover which model(s) appear to best forecast future behavior of the ecosystem. Often the consideration of uncertainty focuses only upon the prediction errors of a single model rather than the credibility of the model structure itself. The credibility of a given model depends on the other models with which it is compared and on the data available. If the data represent only a subset of the potential behavior of the ecosystem, then the model comparison may be biased, or the appropriate model may not even be discovered, because the behaviors of the ecosystem relevant to the appropriate model have not been observed. If an important model is omitted from the set under consideration, substantial errors can occur in assessing credibility, making predictions, and choosing management actions. Thus model uncertainty has critical implications for ecological management, but the assessment of model uncertainty is limited by the range of ecosystem behaviors observed in the data and by the diversity of models created by the analyst. However, it is impossible to consider all possible, or even all plausible, models. Consequently, model uncertainty means that even well-considered management should expect surprises (Peterson, Carpenter, and Brock 2003). But even these standard caveats on the use of models make little mention of the complexities introduced by human interventions.

The future state of human drivers is even more difficult to predict than ecological dynamics due to rapid social change and the reflexive nature of people. Resource management policy is often based on existing technologies and values or the continuation of existing trends. However, new inventions (in the broadest sense of the word) frequently change the way people impact ecosystems. Furthermore, humans are reflexive and therefore consider the consequences of their future behavior before making a decision. This reflexivity can make predictions self-fulfilling or self-negating, which makes predicting future behavior more difficult if not impossible (Carpenter, Ludwig, and Brock 1999).

In adaptive assessment, models are viewed as hypotheses and as such cannot be validated, only invalidated. The models are caricatures of reality, only including what is essential. Therefore what is important is model credibility, not validity. It is only after resisting attempts at invalidation that a model becomes credible. One way of attempting invalidation is to compare the model output with historical data (verified data, not interpreted). Other means of invalidation include trial and error approaches, that is, comparing the model predictions with what happens in the real world. There may be natural trials where model output can be compared to natural experiments. Another way is to compare the behavior of alternative models. Once the models (or sets of models) have resisted invalidation, they can be used to evaluate alternative actions. Walters (1986) concludes that the adaptive assessment phase utilizes models to highlight uncertainties and foster communication, but that they cannot and should not be used to prescribe specific management actions, arguing (1986, 1997) that building models for predictive responses has not been successful because uncertainties in management cannot be resolved by existing scientific approaches and techniques. Consequently, most management decisions are gambles because of the uncertainty of outcomes of a management action.

The previous paragraphs have discussed the role of causal modeling as a framework for processing information during adaptive environmental assessments. The model construction is the important part of the processing, not the actual model or models developed. Indeed Walters (1986) indicates that these models are discarded at the end of the assessment phase, as their utility to summarize and synthesize information is over. However, the synthesis or assessment made during the modeling exercises is critical to 1. deciding upon an agreed set of hypotheses to test, and 2. designing experiments to test both the understanding of system dynamics and the effects of subsequent interventions and policy implementation. The development of hypotheses (or explanations) and subsequent implementation and evaluation of policies derivative of those explanations can be construed as a type of social learning, as described in the next section.

FOSTERING SOCIAL LEARNING THROUGH AEAM

In this chapter we use the phrase *social learning* to describe all the activities involved in AEAM (e.g., moving through the processes depicted in figure 8.1).

In essence, social learning is a structured process by which a group proposes a conceptual representation or scheme and then applies empirical evidence to reject or sustain that scheme. Learning can occur incrementally, episodically, or transformatively (Holling and Gunderson 2002).

Incremental learning occurs as plans, models, and policies are implemented and evaluated. Models or schemes are assumed to be correct, and learning is characterized by collecting data or information to update these models. In bureaucratically dominated resource systems the activity of learning is carried out largely by self-referential professionals or technocrats who primarily view dealing with this type of change and learning as a problem solution (Westley 2002). Passive forms of adaptive management promise this type of learning.

Episodic learning is discontinuous in time and space. It is generated by slow-fast dynamics that reveal the inadequacies of the underlying models or policies. This type of learning occurs after an environmental crisis, where policy failure is undeniable (Gunderson, Holling, and Light 1995). In this case the learning is described as double loop, where the underlying model or scheme is questioned and rejected (Argyris 1977). This is also characterized as problem reformation. In bureaucratic resource systems this type of learning is facilitated by outside groups or charismatic integrators (Blann, Light, and Musumeci 2003). Experimentation through adaptive management can lead to episodic learning, as described later in the Grand Canyon experience.

Transformational learning is characterized by cross-scale surprise and/or the emergence of novel solutions. In these cases learning involves solving problems of identifying problem domains among sets of wicked and complex variables (Westley 2002). Transformational learning involves several levels in a panarchy, not simply one level (Holling and Gunderson 2002). It has also been described as evolutionary learning (Parson and Clark 1995) where not just new models or scheme are developed but also new paradigmatic structures. The development of Everglades restoration in the adaptive assessment process is one example of transformational learning. In this case a number of problem domains (ecological, social, and economic) were solved by viewing restoration as a win-win solution for all sectors and not a zero-sum game of conflict for water among agricultural, urban, and conservation sectors. In the assessment processes a novel set of integrative solutions were developed that changed the paradigm of system management from one of failure to achieve multiple objectives (economic development was successful, but at the cost of environmental conservation) to one where all objectives were attainable (that is, economic development and conservation of resources could be attained).

Each of these forms of social learning involves the creation and development of social networks. Network components may be management institutions, such as governmental agencies, groups such as stakeholders, tribes, or associations, or individuals. The network is comprised of these groups and the way in which these groups interact formally or informally.

For example, resource management in the Everglades is done through a large complicated network comprised of governmental agencies at the federal, state, and local level and nongovernmental agencies that represent environmental and agricultural interests. The Everglades network has a high institutional diversity (numerically and functionally). The network is dominated by governmental agencies and formal policies. As such, it has developed into a formal, closed network. Because of the number of components in the network, it is a very complex management structure. Associated with a complex network are high transaction costs due to the number of components; high coordination is required and a high degree of inertia. All this leads to strong forces to maintain the status quo (Light, Gunderson, and Holling 1995), forces that can be counteracted by informal social networks.

Adaptive assessments are informal and involve the empowerment and expansion of small social networks (Holling and Chambers 1973). These small networks have also been called epistemic communities (Haas 1990) because they generally arise from a scientific or technical group whose focus is on learning. Their power comes from a freedom to question assumptions and to synthesize information with little constraint. Experience has indicated that these assessments require diverse perspectives and a heterogeneous set of knowledgeable participants to work. With these ingredients, new perspectives and integrative understanding can emerge. In the most successful assessments, attitudes that differ among groups are changed into a new shared vision. For example, a set of six workshops held over eighteen months in the early 1990s laid the foundation for the current restoration plan for the Everglades in Florida (Walters, Gunderson, and Holling 1992; Gunderson 1999).

Communication of information is key to building and sustaining these social networks for learning. Holling (1978) and Walters (1986) both emphasize the importance of screening alternative policies and communicating those results to policy makers. A number of techniques including visualization and storytelling are important and effective in engaging a broader community in the results of an integrated model. Examples of such modes, include audiovisual presentations (slide shows, storyboards, computer animations) that attempt to compress the complexities of both the analysis and results into a form that address these key policy alternatives. Holling and Walters recommend that communication start early in the assessment process and that decision makers and the public be actively involved from the beginning, rather than passively informed at the end. There are three arguments made for doing this: that it is a fairer way of making social decisions, that participation of stakeholders in a decision process increases the likelihood that they will accept the outcome, and that people bring important knowledge to an assessment process that will improve its quality. This approach is similar to the concept of the extended peer community advocated by postnormal science (Kay et al. 1999).

Diversity plays a functional role in AEAM processes, as suggested in chapter 2 of this volume. A diverse set of disciplinary participants in the adaptive assessment processes are critical to develop a broad set of alternative explanations of resource dynamics and solutions to resource issues. One of the aims of adaptive assessment is to bring together a range of disciplinary perspectives and experience to seek explanations. For example, hydrologists rarely interact with biologists or social scientists. In AEAM the computer model acts a translator, forcing alternative views into a common language or framework (Walters 1986). Functional groups emerge in the processes of AEAM as well. Holling and Chambers (1973) describe caricatures of individuals or groups with functional roles, such as suppliers of information, wild speculators or dreamers, conservative empiricists, saboteurs, and peerless leaders. The presence of each of these groups is critical to the process of social learning.

In this section we have attempted to describe some of the intersections between adaptive environmental assessment and management as well as the themes of information processing and learning based networks. To further elaborate these linkages, we present two contrasting examples of implementation of AEAM, the Everglades and the Grand Canyon ecosystems.

EXAMPLES OF AEAM AND SOCIAL LEARNING

Adaptive management has been successful in large-scale systems where sufficient ecological resilience, learning networks, and a willingness to experiment exist. Ecological resilience, or the capacity for renewal in a dynamic environment (*sensu* Holling 1973), provides an ecological buffer that protects the system from the failure of management actions that are taken based upon incomplete understanding. This property allows managers to affordably learn and change. In addition to learning networks, trust, cooperation, and other forms of social capital are necessary for implementing management actions designed for learning as much as other social objectives (Gunderson 1999, 2001). Other lessons and recommendations from past applications of AEAM are described in table 8.1.

The concepts and techniques of adaptive assessment and management have been applied to hundreds of resource systems around the world. This does not include the hundreds more or so applications that claim to be adaptive management, but are some variant of "management by objective with irregular updating" and do not focus on addressing key uncertainties of resource issues. While adaptive assessments have been done in many systems, few cases of adaptive management exist. From his experience in riparian systems, Walters (1997) states that out of twenty assessments, only seven resulted in experimental management. He provides reasons for failure (defined as not conducting adaptive—experimental—management) that include 1. the belief that further modeling and monitoring will resolve uncertainties, 2. experimentation is costly and risky, 3. experimental approaches are opposed by special interests, and d. value conflicts among scientists and or stakeholders cannot be resolved. Some of the difficulties in adaptive assessment and management are illustrated in lessons from two regions of the U.S., the Everglades of Florida and the Grand Canyon in northern Arizona.

The Everglades of Florida and the Grand Canyon ecosystem are complex social-ecological systems where undesirable ecosystem shifts (eutrophication, species endangerment, loss of habitat, and biodiversity) have resulted from large-scale water management projects. In both cases the restoration of resilience has been a societal objective. Restoring lost ecological values has been attempted by using large governmental programs that are estimated to cost millions to billions of dollars. In both areas adaptive assessments have identified a set of hypotheses regarding the causes of decreased resilience and what is needed to restore diminished ecosystem functions and services. In both cases the hydrologic regimes are key variables that are being manipulated for restoration. In the Everglades the focus is on restoration of clean water flow to the remnant pieces of the shallow wetland to reverse declines in biotic resources (vegetation patterns, aquatic populations, and

TABLE 8.1. Fostering Social Learning Through Adaptive Assessment and Management

RECOMMENDATION	EXPLANATION
Embrace complexity and change	Managing complex systems requires approaches that understand and manage change, rather than stability and stasis. These approaches should be adaptive and flexible for dynamic systems. Processes should be structured to simplify the complex.
Focus on sources of uncertainty	Seek to partition resource issues into ecological, social, political, and institutional components for analysis. Multiple competing explanations of each component should be generated and tested through modeling and information synthesis. Do not push singular explanations or ignore alternative explanations for political or social reasons.
Design processes to resolve uncertainty	Develop actions to test and winnow uncertainties: in the ecological system through management actions and integrative synthesis; in the social system through resolution of values and attitudes between groups to a new, shared vision. Inaction delays understanding. Create cooperation and transform conflict, but recognize some rhythm of conflict and ensure that channels for expressing dissent and disagreement are always open.
Create a "safe to fail" system	Ecological resilience provides a buffer against failed experiments. If resilience is not present, it should be restored. Actions should be designed that are safe for personal careers, institutions, and agencies. If not, assurance is needed.
Create arenas for discourse	Develop arenas where assumptions are questioned, boundaries are challenged, and limits are relieved. Design flexible and adaptive processes that can be modified, sustained, or suspended.
Develop learning networks	Develop epistemic communities that focus on learning, probing, testing and evaluation. Evaluate and monitor outcomes of past interventions and encourage reflection and then changes in practices. Develop diverse modes of communication between and among individuals, groups, sectors, and agencies.
Seek peerless leaders	Leadership is not about creating and implementing a vision, but about creating and protecting processes designed for learning and creativity. Develop a small team of leaders, deeply knowledgeable and wise in the ecosystem and social system. In the best cases leadership spans social-political scales; one person can do it for a time, but several are better locally, regionally, and politically.

(continued)

TABLE 8.1. (*Continued*)

Scale is important	No single scale in space or time is correct. Multiple assessments should be made that cover multiple scales.
	Time horizon for assessment and management should focus on scales generally ignored, such as the range of multiple decades.
	Recognize cross-scale interactions. Change is both bottom up and top down. Small-scale social processes are creative, but tend to lack power and influence. Large-scale social processes can help, but can also destroy local adaptive capacity.
Invent creative solutions	Develop and maintain a portfolio of projects, waiting for opportunities to open.
	Encourage small-scale tests that challenge understanding, social, and political bounds. Seek to avoid big-scale collapses.
	Seek ways to decrease the cost of learning. Too much fiscal and social capital can be counterproductive and stifle creativity.

nesting/wading bird populations). In the canyon it is the managed flood events (high flows) that are used to move suspended sediments onto bars and beaches as well as to attempt to recover endangered fish populations.

While both systems claim to have ongoing adaptive management programs, the Grand Canyon has an active experimental management program, while the Everglades has an active planning program. The Grand Canyon adaptive management program has conducted two large-scale flow experiments, one in 1996, the other in 2004. These experiments consisted of the release of large volumes of water from the Glen Canyon dam as the primary control variable. The expectations (hypotheses) were tested to see if the floods would deposit suspended sediments onto sandbars and beaches along the river corridor. In conducting these experimental releases, scientists discovered that their conceptual model of sediment storage was wrong. Rather than sediment being stored primarily in bed loads, they discovered a large percentage was stored in eddies within the river. Another set of experiments was developed in 2002 to understand effects of flow, water temperature, and predators on recruitment dynamics of the endangered Humpback chub. The Everglades, however, has been planning management actions for over a decade, with the promise of actions rather than any actual tests of hypotheses. In the early 1990s a series of adaptive assessment workshops were conducted. The conclusion of those workshops was that restoration of lost environmental values was feasible (Walters, Gunderson, and Holling 1992; Davis and Ogden 1994), but would require a concerted effort of integrative solutions applied at the scale of the ecosystem. Since then a number of formal planning processes (U.S. Corps of Engineers Restudy, Governors Commission for a Sustainable South Florida) led to the passage of the Everglades Restoration Act in 2000 by the U.S. Congress. That act authorized up to eight billion dollars for restoration purposes. More planning has occurred, with a few pilot programs and no structured experimentation. The reasons for this difference lie in the social and institutional structures in each system.

The Everglades and the Grand Canyon examples diverge with respect to their ability to cultivate social learning. The Everglades process of experimentation has been trapped by special interest groups (agriculture and environmentalists) that seek certitude in policy, rather than understanding through experimentation. In part due to a long history of conflict, the stakeholders in the Everglades do not have a great deal of trust for one another. While plans for adaptive management actions have been made, they have never been implemented because of resistance by special interest groups who want specific guarantees of water allocations. The Grand Canyon group, on the other hand, has developed an "adaptive management work group" that uses planned management actions and subsequent monitoring data to test hypotheses and build understanding of ecosystem dynamics. This group is characterized by a diverse set of leaders (not a single leader, but multiple, overlapping leadership roles filled by different persons). The leaders in the Grand Canyon understand the uncertainties and complexities of the system and believe that resolution of environmental issues can only be discovered, not determined by predetermined policy. As such, they have not provided an overarching vision, complete with targeted goals and objectives, as much as provided opportunities and windows for experimentation. That is, they have created "space" for experimentation and learning (Gunderson 2003). This has generated a great deal of trust among stakeholders and a more open and flexible institutional setting for dealing with multiple objectives, uncertainty, and the possibility of surprising outcomes. Such emergent governance that creates new institutional platforms for adaptive management is evolving in many places around the world (Folke et al. 2002). The Everglades appears stuck in a "rigidity trap" (Holling and Gunderson 2002) characterized by a tightly controlled, bureaucratic management system that is intent on conserving past conflicts rather than discovering sustainable futures.

Over three decades have elapsed since a few experimental workshops on resource science generated a novel approach to dealing with environmental problems (Holling and Chambers 1973). The application of adaptive environmental assessment and management grew in an era of increasing awareness of the environment and the need for technical analysis of environmental impacts.

The conceptual underpinnings for AEAM are quite simple; there will always be inherent uncertainty in the dynamics and behavior of complex ecologic systems. Although new theories arise, and computers become faster, there is an inherent unpredictability to ecological systems because of multiple stability domains, nonlinear interactions. The genius of adaptive assessment and management lies in the continuing recognition and confrontation of uncertainty. With uncertainty as part of the picture, management actions focus on resilience and flexibility to deal with an uncertain future or with safe-to-fail designs that acknowledge inevitable change and surprise.

The central argument of AEAM is integrative learning. That learning is fostered not by trial and error but by a structured process. Integrative learning is described in phases of assessment, policy as hypotheses, management actions as tests, and evaluation. Assessment phases of AEAM involve bounding issues, focusing on shared understanding of policy-relevant hypotheses and screening alternatives for testing. In some resource assessments a singular hypothesis can be generally agreed upon for testing via management actions. In many cases multiple, competing hypotheses can be generated and spawn a suite of activities; research, modeling, data collection, and synthesis to help sort among these competing ideas.

Having a strong conceptual and theoretical underpinning, adaptive management seems an obvious approach, at least when many managers are desperately seeking alternatives. Experience, however, has indicated mixed success. Adaptive assessments have robustly transformed understanding and generated totally new management schemes in many areas. Active adaptive management (structured to test or sort among competing hypotheses) has failed in most large-scale applications. These failures have been a result of a lack of ecological resilience, inability to control experimentation at appropriate scales, and the lack of flexibility, trust, and openness in the human management system. In cases where adaptive management is precluded, the management objective should be to restore resilience and flexibility in the system.

Adaptive management continues to develop new conceptual and practical tools. Important areas of research areas include improving understanding of cross-scale connections in social-ecological systems, building resilience, and effectively engaging an extended peer community. These research frontiers are related, as increased human transformation of the biosphere and global connections means that the cross-scale connections that shape social-ecological systems influence resilience, both social and ecological. Management actions that increase a system's resilience allows for more experiments to be undertaken. Therefore determining how to incorporate resilience building in management is an important area of ongoing research. Finally, adaptive management requires better ways of building lasting engagements among extended peer communities, to integrate the expertise and energy of people who live in and manipulate ecosystems together with scientific knowledge.

Adaptive assessment and management confronts the inherent uncertainty of resource management systems and in so doing intersects with the themes of this volume. Scale is explicitly defined in the assessment phase by developing an integrated ecosystem model. The model is bounded in both spatial and temporal domains and can only minimally deal with larger, slower processes and smaller, faster processes. This scale bounding is one source of uncertainty in the assessment process. In the management phase, limits in space and time are imposed by the ability of managers to act within a system. For example, in the Colorado River, experiments with flow can be done at the spatial scale of a river segment, because of the ability to manipulate flows from a dam. Larger, more complicated adaptive management experiments have failed because of the inability of humans to exert control over large scales in systems such as the Columbia River or Everglades (Walters 1997). Social networks that are centered on epistemic communities and extend to stakeholder and policy groups are developed in adaptive assessments and are critical to successes and failures of implementation and management (Gunderson 1999). A diverse set of functional roles emerges in adaptive workshops that are also critical to success (Holling and Chambers 1973). Information is processed and evaluated in the assessment and management phases by the use of structured processes designed to learn and innovate. In spite of limitations, adaptive management is a robust approach for dealing with complex natural resource issues; from renewable resource harvest and use to strict preservation. Its focus on learning, integrating understanding and generation of novelty, are as relevant now as thirty years ago. The adaptive management approach can bridge gaps in communication among technical staff, scientists, resource practitioners, and stakeholders. The development of epistemic social networks that link understanding, management, and governance in open and flexible ways will likely help confront the complexities of managed resource systems in the future.

REFERENCES

Argyris, C. 1977. "Double-Loop Learning in Organizations." *Harvard Business Review* 55:115—25.

Barreteau, O., F. Bousquet, and J.-M. Attonaty. 2001. Roleplaying Games for Opening the Black Box of Multi-Agent Systems: Method and Lessons of Its Application to Senegal River Valley Irrigated Systems." *Journal of Artificial Societies and Social Simulation*, vol. 4, no. 2. http://www.soc.surrey.ac.uk/JASSS/4/2/5.html.

Berkes, F., J. Colding, and C. Folke. 2003. Navigating Social-Ecological Systems: Building Resilience for Complexity and Change. Cambridge: Cambridge University Press.

Blann, K., S. Light, and J. Musumeci. 2003. "Facing the Adaptive Challenge." In F. Berkes, J. Colding, and C. Folke, eds., *Navigating Social-Ecological Systems: Building Resilience for Complexity and Change*. Cambridge: Cambridge University Press.

Carpenter, S. R. 2002. "Ecological Futures: Building an Ecology of the Long Now." *Ecology* 83:2069–2083.

Carpenter, S. R., D. Ludwig, and W. A. Brock. 1999. "Management of Eutrophication for Lakes Subject to Potentially Irreversible Change." *Ecological Applications* 9:751–771.

Carpenter, S. R., W. Brock, and P. Hanson. 1999. "Ecological and Social Dynamics in Simple Models of Ecosystem Management." *Conservation Ecology*. http://www.consecol.org/vol3/iss2/art4/.

Clark, J. S., S. R. Carpenter, M. Barber, S. Collins, A. Dobson, J. A. Foley, D. M. Lodge, M. Pascual, R. Pielke, W. Pizer, C. Pringle, W. V. Reid, K. A. Rose, O. Sala, W. H. Schlesinger, D. H. Wall, and D. Wear. 2001. "Ecological Forecasts: An Emerging Imperative." *Science* 293:657–660.

Davis, S. M., and J. C. Ogden. 1994. *The Everglades: The Ecosystem and Its Restoration*. Delray Beach, FL: St. Lucie.

Folke, C., S. R. Carpenter, T. Elmqvist, L. Gunderson, C. S. Holling, and B. Walker. 2002. "Resilience and Sustainable Development: Building Adaptive Capacity in a World of Transformations." *Ambio* 31(5): 437–440.

Folke, C., S. R. Carpenter, B. Walker, M. Scheffer, T. Elmqvist, L. Gunderson, C. S. Holling. 2004. "Regime Shifts, Resilience, and Biodiversity in Ecosystem Management." *Annual Review of Ecology and Systematics* 35:557–581.

Gunderson, L. H. 1999. "Resilience, Flexibility, and Adaptive Management." *Conservation Ecology.* http://www.ecologyandsociety.org/vol3/iss1/art7/.

Gunderson, L. H., 2001. "Managing Surprising Ecosystems in Southern Florida." *Ecological Economics* 37(2001): 371–378.

Gunderson, L. H. 2003. "Adaptive Dancing." In F. Berkes, J. Colding, and C. Folke, eds., *Navigating Social-Ecological Systems: Building Resilience for Complexity and Change.* Cambridge: Cambridge University Press.

Gunderson, L., and C. Holling. 2002. *Panarchy: Understanding Transformations in Human and Natural Systems.* Washington, DC: Island.

Gunderson, L., C. Holling, and S. Light. 1995. *Barriers and Bridges to the Renewal of Ecosystems and Institutions.* New York: Columbia University Press.

Gunderson, L., and L. Pritchard. 2002. *Resilience and the Behavior of Large Scale Systems.* Washington, DC: Island.

Haas, P. 1990. *Saving the Mediterranean: The Politics of International Environmental Cooperation.* New York: Columbia University Press.

Hilborn, R., and M. Mangel. 1997. *The Ecological Detective: Confronting Models with Data.* Princeton: Princeton University Press.

Holling, C. S. 1973. "Resilience and Stability of Ecological Systems." *Annual Reviews in Ecology and Systematics* 4:1–23.

Holling, C. S. 1978. *Adaptive Environmental Assessment and Management.* Caldwell, NJ: Blackburn.

Holling, C. S. 1981. "Forest Insects, Forest Fires, and Resilience: Fire Regimes and Ecosystem Properties." U.S. Forestry Service General Technology reprint WO-26:445–464.

Holling, C. S. 1986. "The Resilience of Terrestrial Ecosystems: Local Surprise and Global Change." In W. C. Clark and R. E. Munn, eds., *Sustainable Development of the Biosphere*, pp. 292–317. Cambridge: Cambridge University Press.

Holling, C. S., and A. D. Chambers. 1973. "Resource Science: The Nurture of an Infant. *Bioscience* 23(1): 13–20.

Holling, C. S., and L. H. Gunderson. 2002. "Resilience and Adaptive Cycles." In L. H. Gunderson and C. S. Holling, eds. *Panarchy: Understanding Transformations in Human and Ecological Systems*, pp. 25–62. Washington, DC: Island.

Holling, C. S., and G. K. Meffe. 1996. "Command and Control and the Pathology of Natural-Resource Management." *Conservation Biology* 10:328–337.

Hughes, T. P., A. H. Baird, D. R. Bellwood, M. Card, S. R. Connolly, C. Folke, R. Grosberg, O. Hoegh-Guldberg, J. B. C. Jackson, J. Kleypas, J. M. Lough, P. Marshall, M. Nyström, S. R. Palumbi, J. M. Pandolfi, B. Rosen, J. Roughgarden. 2003. "Climate Change, Human Impacts, and the Resilience of Coral Reefs." *Science* 301:929–933.

Johnson, B. L. 1999. "The Role of Adaptive Management as an Operational Approach for Resource Management Agencies." *Conservation Ecology.* http://www.consecol.org/vol3/iss2/art8/.

Kay, J. J., H. A. Regier, M. Boyle, and G. Francis. 1999. "An Ecosystem Approach for Sustainability: Addressing the Challenge of Complexity." *Futures* 31:721–742.

Lee, K. 1993. *Compass and Gyroscope: Integrating Science and Politics for the Environment.* Washington, DC: Island.

Light, S., L. Gunderson, and C. S. Holling. 1995. "The Everglades: Evolution of Management in a Turbulent Environment." In L. H. Gunderson, C. S. Holling, S. S. Light, eds., *Barriers and Bridges to the Renewal of Ecosystems and Institutions.* New York: Columbia University Press.

Lynam, T., F. Bousquet, C. Le Page, P. d'Aquino, O. Barreteau, F. Chinembiri, and B. Mombeshora. 2002. "Adapting Science to Adaptive Managers: Spidergrams, Belief Models, and Multi-Agent Systems Modeling." *Conservation Ecology.* http://www.consecol.org/vol5/iss2/art24/.

Ludwig, D., D. Jones, and C. S. Holling. 1978. "Qualitative Analysis of Insect Outbreak Systems: The Spruce Budworm and the Forest." *Journal of Animal Ecology* 47:315–332.

Ludwig, D., M. Mangel, B. Haddad. 2001. "Ecology, Conservation, and Public Policy." *Annual Review of Ecological Systems* 32:481–517.

Millennium Ecosystem Assessment. 2003. *Ecosystems and Human Well-being : A Framework for Assessment.* Washington, DC: Island.

National Research Council. 1999. *Our Common Journey: A Transition Toward Sustainability.* Washington, DC: National Academy.

Ostrom, E. 1990. *Governing the Commons: The Evolution of Institutions for Collective Action.* New York: Cambridge University Press.

Parson, T., and W. C. Clark. 1995. "Evolutionary Learning." In L. H. Gunderson, C. S. Holling, and S. Light, eds., *Barriers and Bridges to the Renewal of Ecosystems and Institutions.* New York: Columbia University Press.

Peterson, G. D. 2002. "Forest Dynamics in the Southeastern United States: Managing Multiple Stable States." In L. Gunderson and L. Pritchard Jr., eds., *Resilience and the Behavior of Large Scale Ecosystems*, pp. 227–246. Washington, DC: Island.

Peterson, G. D., D. Beard, B. Beisner, E. Bennett, S. Carpenter, G. Cumming, L. Dent, and T. Havlicek. 2003. "Assessing Future Ecosystem Services: A Case

Study of the Northern Highland Lake District, Wisconsin." *Conservation Ecology.* http://www.consecol.org/vol7/iss3/art1/.

Peterson, G. D., S. R. Carpenter, and W. A. Brock. 2003. "Model Uncertainty and the Management of Multi-state Ecosystems: A Rational Route to Collapse." *Ecology* 84:1403–1411.

Peterson, G. D., G. S. Cumming, and S. R. Carpenter. 2003. Scenario Planning: a Tool for Conservation in an Uncertain World." *Conservation Biology* 17:358–366.

Raskin, P., G. Gallopin, P. Gutman, A. Hammond, and R. Swart. 1998. *Bending the Curve: Toward Global Sustainability.* PoleStar Series 8. Stockholm: Stockholm Environment Institute.

Rittel, H., and M. Webber. 1973. "Dilemmas in a General Theory of Planning." *Policy Sciences* 4:155–169.

Röling, N. G., and M. A. E. Wagemakers. 1998. "Facilitating Sustainable Agriculture: Participatory Learning and Adaptive Management in Times of Environmental Uncertainty." New York: Cambridge University.

Scheffer, M. 1998. "The Ecology of Shallow Lakes." London: Chapman and Hall.

Scheffer, M., S. R. Carpenter, J. Foley, C. Folke, and B. Walker. 2001. "Catastrophic Shifts in Ecosystems." *Nature* 413:591–696.

Steffen, W., A. Sanderson, P. D. Tyson, J. Jäger, B. Moore III, P. A. Matson, K. Richardson, F. Oldfield, H.-J. Schellnhuber, B. L. Turner II, R. J. Wasson. 2004. *Global Change and the Earth System: A Planet Under Pressure.* New York: Springer.

Vitousek, P. 1997. "Human Domination of Earth Ecosystems." *Science* 277(5325): 494–499.

Walker, B. H., D. Ludwig, C. S. Holling, and R. M. Peterman. 1981. "Stability of Semiarid Savanna Grazing Systems." *Journal of Ecology* 69:473–498.

Walters, C. 1986. *Adaptive Management of Renewable Resources.* New York: McGraw Hill.

Walters, C. 1997. "Challenges in Adaptive Management of Riparian and Coastal Ecosystems." *Ecological Applications.* http://www.consecol.org/vol1/iss2/art1.

Walters, C., L. Gunderson, and C. S. Holling. 1992. "Experimental Policies for Water Management in the Everglades." *Conservation Biology* 2:189–202.

Walters, C., and C. S. Holling. 1990. "Large-scale Management Experiments and Learning by Doing." *Ecology* 71:2060–2068.

Walters, C., J. Korman, L. E. Stevens, and B. Gold. 2000. "Ecosystem Modeling for Evaluation of Adaptive Management Policies in the Grand Canyon." *Conservation Ecology.* http://www.consecol.org/vol4/iss2/art1/.

Westley, F. 1995. "Governing Design: The Management of Social Systems and Ecosystems Management." In L. Gunderson, C. S. Holling, and S. S. Light, eds., *Barriers and Bridges to the Renewal of Ecosystems and Institutions,* pp. 489–532. New York: Columbia University Press.

Westley, F. 2002. "The Devil Is in the Dynamic." In L. Gunderson and C. Holling, eds., *Panarchy: Understanding Transformations in Human and Natural Systems*, pp. 333–360. Washington, DC: Island.

Westley, F., S. R. Carpenter, W. A. Brock, C. S. Holling, and L. H. Gunderson. 2002. "Why Systems of People and Nature Are Not Just Social and Ecological Systems." In L. Gunderson and C. Holling, eds., *Panarchy: Understanding Transformations in Human and Natural Systems*, pp 103–119. Washington, DC: Island.

9

SCALE AND COMPLEX SYSTEMS

Graeme S. Cumming and Jon Norberg

NEARLY ALL PEOPLE are elements of a larger system and as a result are faced with the challenges of understanding what comprises the system they are part of, where they fit in the system, and what they can measure or change about it. The process by which we learn about the world has two scale-dependent components; the actual scales at which patterns and processes occur and the scales at which we obtain data about them. If we are to obtain reliable information about the world, it is crucial that we undertake measurements at the appropriate scales to match the processes we are interested in exploring.

Patterns and processes of interest in ecology, sociology, and economics are usually bounded entitites that exist at particular spatial and temporal dimensions. Complex adaptive systems (CASs) are distinguished not only by their diversity of components, nonlinear behaviors, complex (typically hierarchical) organization, multiscale nature, and homeostatic feedbacks; they are also unique in their ability to self-organize, or adapt, in response to environmental demands. This ability is often determined by cross-scale interactions, as in the transfer of knowledge from local to regional scales or the colonization of localities by members of the regional species pool. When we study a CAS, it is useful to be able to consistently say what is a part of the system and what is not, even if membership in the set of system components is fluctuating or fuzzy. If we cannot provide a system definition then we run the risk of continually trying to track a moving target. Determining the system boundary typically involves assessing system properties across a range of scales to determine which scales are of greatest interest (Pickett and Cadenasso 2002; Cumming and Collier 2005). Boundary conditions specify the largest area or time period at which the system exists.

Patterns and processes that exist outside this boundary are termed the system's context or environment or are simply labeled external factors. In addition to determining boundaries, the choice of scale for a system definition also excludes processes that are at too fine a scale to be of interest. For example, in a study of sustainable forestry practices in Brazil the movements of the individual atoms that make up the community leaders would usually be excluded from active consideration. These movements may be integral to the composition of the system, but they are too far removed from the focal question to be of interest; their scale of variation is so small that it has no influence on variations in the study system.

When considering linked social-ecological systems, the phenomena in which we are interested usually occur across a range of scales. Their magnitudes or topologies may be proportional to scale (linear, isometric, allometric, or self-similar scaling), may be strongest at certain scales and weakest at others (nonlinear scaling), or may be largely independent of scale (fractal or scale free). As Simon Levin (1992, 1999) points out, many of the properties of CASs change predictably with scale. For example, feedback loops involving processes such as natural selection or homeostasis tend to be tighter at small scales and hence to have more direct and more immediate effects. One of the outcomes of small-scale feedbacks may be to create subsets or blocks that function as cohesive units within larger systems. This view parallels the suggestions of Holling (1992) that in ecosystems processes typically occur discontinuously at one or more spatiotemporal scales and that this discontinuity of processes contributes to a pattern in which system attributes or constituents are clustered at particular scales.

When we study a CAS, we are forced to measure and try to understand scale-related changes in the magnitudes of processes and the characteristics of patterns (Wiens 1989; Bellehumeur and Legendre 1998; Dungan et al. 2002). Consequently, we will always need to have some idea of whether our scale of measurement is appropriate relative to the questions that we are asking and the CAS that we are studying. Good science will be facilitated by clarity in defining/determining our scales of interest, the scale-dependent ways that we obtain data, and being able to place what those data tell us in an appropriate context.

As previous chapters have made clear, systems are made up of interacting units. The distinction between system and unit is one of scale; at a finer scale of analysis units may be systems in their own right. Units and their relationships, or linkages, form networks that span space and change through time. The types of units and interactions within the system, as well as the responses of units to interactions, determine how units and groups of units depend on one other. Nested hierarchies can be seen as emergent properties of the complex behaviors of interacting units. The dynamics of a hierarchical complex system emerge partly from the particularities, locations, and diversity of units involved, as analyzed in chapters 1 and 2. Consideration of a CAS as a network of interacting nodes or agents may give a good schematic view of scale-related issues (chapter 3 and 4). For example, trophic interactions between organisms can be described as a food web, which is a network. The food web describes the strengths of interactions between species in both time and space and can be used to analyze the degree of compartmentalization in spatiotemporal dimensions as well as the organizational structure of the community (Polis and Strong 1996; Polis, Power, and Huxel 2004). Moreover, the responses of units to external drivers and their interactions with other units (i.e., information processing) are crucial to system function. These topics are considered in detail in chapters 5 and 6. Cross-scale interactions are among the most important whole-system dynamics with regard to large-scale changes (Holling 1973; Peterson, Allen, and Holling 1998; Scheffer et al. 2001), but are notoriously hard to detect from within the system itself.

In this chapter we offer a summary of current ideas about scale and scaling as they relate to the study of CASs. After considering multidisciplinary applications of scale and related concepts, we discuss some of the approaches that have been used in the field of scaling. Last, we reflect on scale and scaling as portrayed throughout this volume and consider some of the more novel, scale-related ideas that have emerged during the process of writing it.

DEFINING SCALE

In the biophysical sciences scale is generally defined as a property of the dimensions of space and time. Spatiotemporal scale has two main components, grain and extent (Turner, Gardner, and O'Neill 2001). Grain refers to the resolution of analysis, and extent to the coverage. For instance, an image of a house might be high resolution (fine grain) or low resolution (coarse grain) while maintaining the same extent. The finer the grain, the easier it is to pick out small differences and precise features of interest in the picture. Coarser-grained images will look increasingly blurred to the human eye, but may be useful for understanding dominant features or overall color composition. Alternatively, two photographs could be taken at the same resolution of two different objects, such as a house and a flower. The grain of these images remains the same, but the extent differs; a house is considerably larger than a flower. Grain and extent often vary together because of limitations on data collection, storage, and processing; most high-resolution images have small extents, because a high-resolution image of a very large area would require more memory (i.e., would contain more information) than a computer disk has available. Obviously, although we have illustrated these ideas with spatial scale, the same is true for temporal scale; it is possible to sample very frequently or very seldom (grain), over a very short time period or a very long one (extent).

The distribution or dispersion of patterns or processes of interest within an area or time period (i.e., whether clustered, even, or random and whether rare or common, frequent or infrequent) is partly captured by grain and extent, but is sometimes considered a third component of scale. When making measurements, the average distance between unit objects or processes is termed the sampling interval (Legendre and Legendre 1998). The sampling interval is a key component of kind of information that is obtained; it is possible to sample very frequently or very seldom in either space or time at a constant grain and over a constant extent.

In the social sciences scale assumes an additional meaning: that of institutional size, representation, and power (Gibson, Ostrom, and Ahn 2000). Broad-scale sociological processes are defined more by the large numbers of people involved than by their absolute spatiotemporal dimensions (although the two will often be aligned). Institutions can lead large numbers of people to act in a similar way without any obvious change in either the number or the size of participating entities. Such coordinated actions can produce emergent properties in much the same way that ant colonies function as a unit at a larger scale than the individual. Consideration of social systems in this light identifies an issue that biophysical definitions ignore, the question of connectivity or interaction. Two systems that exist at the same spatial and temporal scales might still differ significantly in terms of the amounts of interaction between individual components (Bodin and Norberg 2005).

CROSS-SCALE INTERACTIONS

System definition at a single scale is relatively straightforward. It is a lot harder to obtain a multiscale system definition that captures scaling relationships between different system components. Patterns and processes that occur at smaller scales are often contained within those at larger scales, thereby forming a nested hierarchy. One of the earliest attempts to describe interactions of entities at different scales introduced the concept of holons, first presented by Koestler (1967) in his book *The Ghost in the Machine*. Holons refer to the units that make up a hierarchy (such as species or individuals in an organization or counties in a state). The original description of holons by Koestler (1967) draws on an analogy to the faces of Janus. Janus was the Roman god of gates and doors (*ianuae*) and of beginnings and endings. Janus was represented with two faces, one that looked forward and one that looked back. Hierachy theory likens holons to Janus the two-faced, arguing that higher holons constrain lower holons and lower holons are the constituents of higher holons (Salthe 1985). While hierarchy theorists made a scientific attempt to define scales by relative changes in process rates (O'Neill et al. 1986), the later concept of panarchy (Holling 2001; Gunderson and Holling 2002) used the concepts of system controls and system mechanisms, arising together from constituents that were adapted to the dynamics and processes observed in a CAS, to address the fundamental problem of cross-scale interactions.

Hierarchy theory has proved to be an essential conceptual tool for thinking about multiscale systems (Simon 1962; Allen and Starr 1982; O'Neill et al. 1986; Wu and Loucks 1995). Hierarchy theory describes hierarchical complex systems as a partially ordered set; system components belong to a group in which members can be described by their relative properties, which may be equal, in addition to whatever commonality (location, function, etc.) led to them being included in the same system in the first place. For example, the set of administrative districts in the USA includes the whole country, individual states, counties, and municipalities. This set is hierarchical because its members can be partly ordered on the basis of their sizes and locations; the U.S. includes Florida, Florida includes Alachua and Broward counties (among others), and Alachua includes municipalities such as those of Gainesville and Waldo. The ordering is termed partial because within a hierarchical level different counties are effectively interchangeable. Additionally, O'Neill et al. (1986) have argued that nested levels can be identified by the relative strength of processes occurring within as compared to between scales. Although not all hierarchies are nested (food webs and chains of command are hierarchical, but levels are not composed of one another directly), many of the same principles that apply to nested hierarchies are also relevant to non-nested hierarchies.

Hierarchy theory offers a way of visualizing and arranging system components in CASs. In the context of scaling its most important generalization is the idea that there are predictable relationships between system components that occur at different levels. These can be summarized by the aphorism "upper levels constrain, lower levels explain." If we are interested in an entity that occurs at a particular hierarchical level, then constraints on its properties (such as their extent, magnitude, or rate) will often come from the hierarchical level above. For example, Alachua county can grow no larger than the state of Florida; and more immediate limits on its growth are imposed by bureaucratic processes and decisions that occur at a state-wide scale. By contrast, the mechanisms that explain the behavior of the level of interest tend to come from lower in the hierarchy. The municipalities in Alachua county make local decisions about roads and protected areas, and these decisions (together with externally imposed constraints such as the amount of funding received from the state government) explain a large amount of the variation in both the pattern and the process of development in Alachua. At certain times, however, these relationships may be inverted and lead to cross-scale impacts, as occurs when spruce budworms build up to sufficient numbers to influence boreal forests (Holling, Gunderson, and Peterson 2002).

Interactions between levels in a hierarchy are examples of cross-scale processes. Although many interactions occur between entities that exist at the same scales, such as individuals buying and selling produce in the Gainesville farmer's market, it is often the cross-scale interactions that are both the most important for overall system dynamics and the hardest to study. Cross-scale interactions may occur via any of a number of currencies, including such things as information, power, laws and regulations, energy, nutrients, and money. In most hierarchies cross-scale interactions may be top down or bottom up. Bottom-up interactions typically occur by contagion or domination. For example, the cumulative sum of individual votes determines the composition of the local government; the cumulative effects of individual trees being cut can lead to forest fragmentation effects that have influences at much broader scales than that of a single tree. Top-down interactions typically occur by control or regulation. They are common in social systems and markets, for example where a federal court determines the eventual outcome of an election.

Cross-scale interactions have a large number of consequences for the dynamics of CASs. They are particularly important in situations where the potential for feedbacks exists. Feedbacks occur when an effect influences its cause. Negative or homeostatic feedbacks exert control over a process by pushing it back toward an equilibrium, such as sweating to control overheating of the human body. Positive feedbacks lead to self-amplifying processes, such as the greenhouse effect, and may result in substantial system change. Positive or negative feedbacks at the same scale are less likely to have systemwide consequences than feedbacks that occur across scales, which can result in the reorganization of hierarchical structures.

STUDYING SCALE-DEPENDENT PATTERNS AND PROCESSES

The basis for scaling studies is to arrange system components, relationships, patterns, or processes according to their scale(s) of action and then try to understand how scale-dependent differences relate to system properties and dynamics. Although the similarities between different approaches to scaling studies are seldom discussed, they can be arranged in a sequence of increasing complexity that runs from the most basic, direct kinds of scaling function through allometric relationships, deviations from normality, fractals, methods that take location into account more explicitly, methods that take system composition into account, methods that take system composition and configuration into account, and finally methods that consider both spatial and temporal variation in the same analysis. We next present a brief summary of each of these incremental developments in scaling theory.

BASIC DESCRIPTIVE FUNCTIONS   The most basic kind of scaling analysis is to simply plot changes in a quantity of interest as a function of scale. For example, species-area curves are among ecology's best-known and most general scaling relationships (e.g., Gleason 1922; Lawton 1999; Scheiner et al. 2000). As the area being sampled increases, the number of species in the area increases. This relationship can usually be fitted to a logarithmic function of the form $S = cAz$, where S is the number of species, A is the area being sampled, c is a constant that describes the starting point of the curve, and z is a constant that describes the gradient (steepness) of the curve. Another example of a simple scaling analysis is the relationship between latitude and species richness, with more species being found closer to the equator in many different taxonomic groups (e.g., Pianka 1966; Gaston 2000).

ALLOMETRY   A common form of scaling analysis is found in allometric and isometric scaling studies. Allometric scaling analyses build on simple scaling functions. However, rather than considering changes in a single quantity with scale, inferences are drawn by analyzing scale-related changes across a range of systems that are presumed to obey the same mechanisms. Many useful inferences can be drawn from consideration of changes in morphometry in relation to body mass across a range of similar organisms (Peters 1983). For example, the body mass of ungulates is an excellent predictor of the area of the underside of the foot (Cumming and Cumming 2003; figure 9.1). Depending on the nature of the inferences that are to be drawn, such scaling studies may need to correct for historical processes; similar organisms are systems that have evolved from a common ancestor, and their similarities may be due to relatedness rather than to current utility.

Classical allometric scaling studies have been revisited in recent times to incorporate a wider range of relationships. Further research in this area has occurred in two main foci: understanding the generality of mechanisms that give rise to scaling patterns and considering other organizational units in addition to individual organisms. So, for example, West, Brown, and Enquist (1997) have proposed that the branching patterns of transport tubes (such as tracheae and phloem vessels) can be predicted from the first principles of surficial exchanges; Brown (1995) has shown that scaling

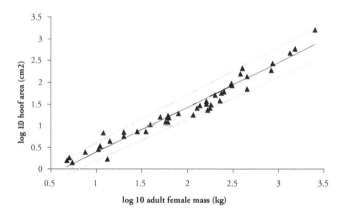

FIGURE 9.1. Plot of the scaling relationship between female body mass and hoof area for forty-five species of African ungulates. The reduced major axis regression line has a Pearson's *r* value of 0.997 and a slope of 1, indicating that the pressure per unit area exerted by ungulates is independent of body mass. The gray lines indicate 90 percent confidence intervals (after Cumming and Cumming 2003).

relationships exist between population and community-level patterns such as the range size of an organism and its relative abundance. As discussed in chapter 2, scaling relationships are used in both cases to argue in favor of generalities and to identify exceptions to general principles. Similar approaches have also been applied in the social sciences (e.g., Longley, Batty, and Shepherd 1991).

BODY SIZE DISTRIBUTIONS   Allometric analyses typically look for the common pattern between systems. Scaling relationships have also been used in a slightly different way, as an assay of system architecture. Holling (1992) proposed that body size distributions (and other phenotypic characters that were under selective pressure) could be used to demonstrate the scale-dependency of ecosystem processes, a hypothesis that he termed *textural discontinuity*. He argued that if animals respond to ecosystem processes at particular scales, then textural discontinuities in the environment should create scale-dependencies in resources, making life at some scales more tenable than at others. After some initial arguments about the best way to quantify discontinuities in body size distributions, the presence of what Holling termed *lumps* and *gaps* in body size distributions has since been supported by a number of studies (e.g., Allen, Forys, and Holling 1999; Marquet and Cofre 1999; Havlicek and Carpenter 2001). The question of whether this support is also support for the textural discontinuity hypothesis is still open; as in standard scaling studies, historical processes must be considered, and a competing (although not exclusive) alternative is that the observed discontinuities arise as a consequence of the step-by-step process of evolution (Cumming and Havlicek 2002). Regardless of how size classes arise, size distributions can have important consequences for ecosystems, social systems, and economies (e.g., Berry 1961; LaBarbera 1989; Meredith 2003).

STOMMEL DIAGRAMS   Most scaling studies look at patterns in either space or time or in a single attribute (such as mass or metabolic rate) that can be compared across multiple systems. A further refinement of the basic scaling approach is to separate out system responses in space and in time. Space-time diagrams were first used by Stommel (1963) to demonstrate the difficulties of accounting for the spatial and temporal variation in fish populations from trawler data. Stommel diagrams, as they have since been called, have been used to describe the dynamics of ecological, social, climatic, and economic systems (figure 9.2, Levin 1992; Wu 1999; Holling, Gunderson,

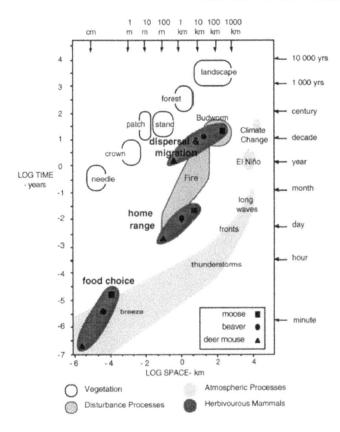

FIGURE 9.2. A Stommel diagram depicting the spatial and temporal scales at which patterns and processes occur in the boreal forest, together with the dominant scales at which three different mammals experience the same environment (reproduced with permission from Peterson, Allen, and Holling 1998).

and Peterson 2002; Westley et al. 2002; Chave and Levin 2003). However, Stommel diagrams have remained primarily descriptive rather than analytical tools. The quantitative comparison of spatial and temporal variation presented in chapter 2 can be seen as a logical progression from Stommel's diagrammatic representation of space and time.

SEMIVARIOGRAMS   Semivariograms and their time series equivalent, autocorrelation analysis, offer a way of tracking the dependency of process magnitudes on spatiotemporal location (Isaaks and Srivastava 1989; Shumway and Stoffer 2000). By comparing the magnitudes of the correlations or semivariances of multiple pairs of measurements of a single variable to

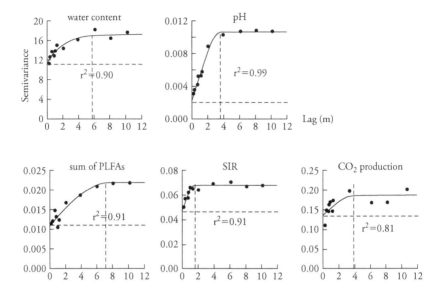

FIGURE 9.3. An example in which semivariograms are used to analyze the spatial scale of different soil properties in a mixed spruce/birch forest. PLFA: Phospholipid fatty acids, a measure of bacterial abundance; SIR: soil induced respiration, a measure of biological activity in the soil (adapted from Saetre 1999).

their proximity to one another, variograms and related techniques allow the description of the scales at which processes vary (figure 9.3). Although semivariograms are extremely useful in some contexts, such as mining engineering, they have not found wide use in either ecology or sociology. This is partly a consequence of the ambiguity of many variograms, which often do not show a clear sill or range and may be difficult to interpret objectively. The kind of categorization adopted by many ecologists, where the world (as viewed through a vegetation map or minimum convex polygon) is seen as consisting of patches and matrix, also mitigates against the application of semivariogram analysis which is more appropriate to data sets that vary continuously through space or time.

FRACTAL GEOMETRY Another approach to scaling problems that has received considerable attention in recent years is fractal geometry (Mandelbrot 1983). Fractals describe the geometry of surfaces that are infinitely rough; measurements of fractal surfaces at smaller and smaller dimensions do not approximate a smooth curve, but instead remain rough. The classic example is that of a coastline. Estimates of the length of the coast will differ depending on the grain at which the coastline is measured, with increasingly smaller grains of measurement picking up roughness at increasingly smaller scales.

Most fractals that have been discussed in the literature, such as the Mandelbrot set or the Julia set, are self-similar at multiple scales. Small pieces of the set look much like the entire set. Although some natural patterns are self-similar, many are not. The boxplot method that is often used to empirically estimate the fractal dimension of an object can be informative for scaling studies, and offers an approach to defining non-linear scaling functions (i.e., that do not assume self-similarity across scales). The boxplot method was first applied by Tarboton, Bras, and Rodriguez-Iturbe (1988) to estimate the fractal dimensions of streams. If a grid of cells is layed over a river network, it is simple to score the numbers of cells that contain stream fragments and those that do not. Changing the box size leads to a different score. Plotting the number of occupied cells against the dimension of a single box produces a function that describes how the space-filling properties of the pattern under consideration change with scale. For streams, this function is typically linear; and the slope of the fitted line is the fractal dimension of the object. However, it is also possible to identify patterns that follow a nonlinear scaling function when measured in this way, creating the possibility for a cross-scale measure similar to a fractal dimension but specified by a polynomial or logarithmic function rather than a single constant.

It has been suggested that particular fractal dimensions in patch or network boundaries may indicate self-organized criticality and potential for sustainability. This proposal is based on the idea that the space-filling properties of edges are underpinned by a set of edge-generating mechanisms that may be of central importance in system function. For example, the boundaries of cities may show predictable changes in their fractal dimensions during the process of urban expansion and (more speculatively) during times of collapse (White and Engelen 1993; Andersson, Rasmussen, and White 2002).

Mark Ritchie and coworkers (Ritchie 1998; Ritchie and Olff 1999; Haskell, Ritchie, and Olff 2002) have applied fractal analysis in a different manner to understanding resource distributions, animal body sizes, foraging patterns, and their relationship to community composition. They argue that if resources are distributed at different scales across the landscape in a fractal manner, and food encounter rates vary according to the body sizes and morphology of individual organisms, then species that share resources will exhibit similar body sizes. If species that exploit a similar niche are to avoid direct competition, they must theoretically be different along at least one niche axis. Applying both "packing" rules and competitive exclusion to the number of species that can coexist in a given landscape off a given set of resources yields some relationships that mirror those found in the real world, such as diversity-area relationships and diversity-habitat fragmentation relationships. The models also predict that animals with larger home ranges should be more sensitive to fragmentation effects.

LANDSCAPE PATTERNS   A multiscale approach to the analysis of pattern scaling has been proposed in the landscape ecology literature, most notably by Jianquo Wu and colleagues (e.g., Wu and Loucks 1995; Wu 1999; Wu et al. 2000). Many landscape ecology studies report single measures of landscape metrics that are considered to be relevant in some way to processes of interest, such as dispersal and edge effects. Wu et al. (2000) have suggested that it would be better to measure landscape pattern across a range of different grain sizes and/or extents in order to provide an indication of the scale-dependency of the results. In other words, rather than selecting a single scale of analysis and treating it as the "correct" scale, a better approach is to integrate across different scales as a way of reducing uncertainty about scale-dependencies. This approach makes excellent sense in theory, but it can be difficult in practice to obtain sufficiently fine-grained and extensive data on ecological system attributes of interest (such as population size or community composition). Despite its operational difficulties, however, the principle that scale-related decisions made in individual studies need to be put into context by considering their location along a scale-dependent continuum of results (and ensuing conclusions) is extremely important.

SCALING SPATIAL AND TEMPORAL VARIATION IN THE SAME ANALYSIS
The simultaneous consideration of both spatial and temporal variation across multiple scales is a new and emerging area of research in scaling studies. Although many community ecological studies have considered both spatial and temporal variation (e.g., Salmaso 1996; Espinosa-Fuentes and Flores-Coto 2004), few of these studies have focused simultaneously on scale. One set of approaches to combining spatial and temporal variation in scaling studies has been developed by Pierre Legendre and colleagues (see

Legendre and Legendre 1998; Borcard et al. 2004). Another approach is presented in chapter 2.

STATISTICAL TREATMENTS OF SCALE   One of the central reasons ecologists have been interested in scale is that it is integral to answering questions of how best to sample in space and time to achieve an adequate representation of interactions or dynamics of interest. For example, studies of the population dynamics of boreal forests must necessarily encompass long time scales and relatively large areas; while studies of bacterial communities in ponds require more frequent sampling at much smaller scales. In general, animals will interact with the world at the spatial and temporal scales at which they perceive it (Holling 1992).

Hurlbert (1984) highlighted a common problem in ecological studies; the use of one scale as a replicate for another scale. For example, when studying the effect of biodiversity on ecosystem functioning, it is not enough to contrast experimental plant communities composed of subsets of the same ten species or different numbers of each of ten species. These experiments will only say something about these particular species assemblages; the replication is on the level of individuals, not communities. In order to test for the general effect of diversity, the diversity treatment has to be replicated with different species, not different individuals of the same species (Huston 1997).

The question of statistical power is ever present in many ecological studies that have focused on processes with strong scale dependencies. Studies that look for either spatial or temporal effects tend to find them, but few studies consider both space and time; sampling is typically in a few locations over a long time period or in many time locations over a short period. If fifty locations have been sampled in space at one or two times, it comes as no surprise if spatial effects are interpreted as being more important than temporal effects. This problem is particularly accentuated in studies that use remotely sensed imagery where each additional image adds considerable time and effort in processing and classification; the spatial patterns of thousands of individual pixels can be compared easily, but many studies have used only two or three different time steps to look at change trajectories. Community ecologists have often perceived spatial and temporal influences as nuisance variables that must be removed from an analysis to detect interactions of interest, such as the effects of nitrogen levels on plant communities. It is only recently that spatial and temporal effects have started to come more into focus as important variables in their own right.

It is clear from this summary that considerable research is still needed in measuring and analyzing scaling effects. We need to study scale as an entity, including both space and time, that explains community structural variation in social, economic, and ecological systems. Scaling studies will initially need to collect large amounts of data and will thus be expensive and long term. As general principles emerge it should then be possible to undertake more focused studies and to more readily interpret past and current studies by placing them in the context of a broader theory of scaling.

SCALE-RELATED THEMES FROM PREVIOUS CHAPTERS

This volume is organized around the central themes of 1. asymmetry (systematic heterogeneity) and diversity, 2. networks, 3. information processing, and 4. applications and overarching ideas. Scale and scaling relationships emerge as important themes in each of these sections. We next revisit the themes of the book, from the explicit perspective of scale.

ASYMMETRY, DIVERSITY, AND SCALE

Chapter 2 focuses on the importance of multiscale heterogeneity and pattern-process interactions as drivers of processes in complex systems. Embodied in the concept of asymmetry is the notion that it is not just heterogeneity that we should care about, but also its arrangement in space, in time, and across different scales. This is particularly true in systems where nonlinear changes in scale-related system properties can occur. For example, neutral landscape models show that if pieces of forest are randomly removed from a landscape, a threshold usually occurs at which the net connectivity of forested areas drops rapidly (Stauffer 1985; Turner, Gardner, and O'Neill 2001). In reality, however, forest fragmentation is seldom random. The location of fragmentation events has a large impact on the net consequences of fragmentation for organisms and people who live in the landscape. The challenge is to develop a mechanistic understanding that incorporates not only dependencies that change with scale but also dependencies that change with location and study system (e.g., With, Gardner, and Turner 1997).

Anisotropy, or a lack of evenness in spatial and temporal distributions, is one of the primary reasons why scaling principles developed in one landscape or for one system are often not transferable to another. For example, a process like primary production may show anisotropy if it changes in rate or magnitude with location.

The conceptual marriage of spatial and temporal variation is another of the central themes of chapter 2. The discovery that spatial and temporal variation are correlated in the Florida landscape, and the possibility that spatial and temporal patterns may influence one another via some kind of feedback, suggest that some rethinking of traditional statistical and methodological approaches to measuring and describing space-time interactions is needed. It is unclear what the sampling demands are for a data set that would allow spatiotemporal feedbacks to be detected in a given system, particularly if the interaction is slow or very dispersed. If a space-time feedback cannot be detected by traditional statistical methods, new analytical developments along the lines of those presented in chapter 2 will be necessary.

Many of the central questions about scale in complex adaptive systems are related to pattern-process interactions. Our measurements of both pattern and process change with scale, and the ways in which they do so are often unexpected. Some processes, such as carbon or nitrogen fixation, are widely dispersed in a relatively isotropic manner at multiple scales and so can be averaged across a landscape. Other kinds of process, such as dispersal, are highly dependent on the specific area and arrangement of components of the landscape. Linear features such as rivers and roads may serve as conduits and pathways that constrain nutrients, people, and water to a particular way of using a landscape. Some environments may be net exporters of nutrients, people, or organisms ("source" habitats) while others are net importers ("sink" habitats). Dependencies between different parts of the system can lead to subsidy effects, whereby one area or component of the landscape continues to function only because it is continually being provided with inputs from another part of the landscape. Most of these kinds of pattern-process relationship are scale-dependent, giving different signals and assuming differing levels of importance at different scales.

Diversity is a dynamic variable, and its growth and decline are affected by a wide range of different processes (Norberg et al., chapter 2). In ecological systems the processes that sustain local diversity are embedded in the regional diversity and the network of dispersal or movement patterns that enable new species or genotypes to immigrate. The primary natural processes driving declines in diversity (i.e., the selective process) are environmental perturbations and competitive exclusion. Perturbations may both facilitate and reduce diversity. Environmental change can remove species from the system. However, local temporal variability in environmental drivers can also act as a mechanism that sustains local diversity by reducing the effects of competition and altering dominance relationships. This process has been well documented in fire-generated patch mosaics, where some early colonizers and grasses can persist in the system only if fire is maintained (e.g., Mbow, Nielsen, and Rasmussen 2000). There has been a considerable amount of work on the scaling of perturbations and the responses of ecological communities to particular kinds of perturbation. In 1978 Joseph Connell considered a number of possible mechanisms by which disturbances might influence diversity and argued in favor of the intermediate disturbance hypothesis, which suggests that the relationship between diversity and disturbance is an inverted U-shape; diversity is highest when disturbances occur at an intermediate magnitude, intensity, or frequency (Connell 1978). It has since been shown that the influence of disturbance on diversity is also affected by system productivity (Huston 1994; Kondoh 2001) and by scale (Waide et al. 1999), since more productive systems recover faster and may be affected in different ways by the same disturbance, particularly in a situation where population or community persistence is threshold driven.

Species-area relationships, as discussed in the earlier section, only describe a static pattern. The important underlying processes of dispersal and extinction, which cause the patterns of diversity, have not been given the same attention in terms of analyzing their scale dependencies (e.g., Chave, Muller-Landau, and Levin 2002). Hubbell (2001), in *The Unified Neutral Theory of Biodiversity and Biogeography*, proposes a framework that links process to community-level patterns but does not fully explore the consequences of his model for scaling relationships.

The relationship of diversity to scale can be separated conceptually into two different arenas, the internal diversity of the complex adaptive system and the features of the environment in which the CAS operates. Where a CAS has existed for a long time in a given environmental context (context here refers to both the biophysical environment and the spatiotemporal heterogeneity of selective processes acting on the system), the system will reflect its environment in its composition. In an ecological context the species that are present reflect historical levels and variations in factors that constrain interspecific competition. If gradual changes in constraining factors occur, such as increases in the variability of precipitation, the capacity of the CAS to adapt and respond to change will be determined by its current diversity and its ability to generate new diversity.

NETWORKS AND SCALE

Webb and Bodin (chapter 3, this volume) introduce the concept of scale in networks and how it affects network dynamics. In networks the numbers of nodes (agents, locations, or other interrelated entities) and edges (connections between nodes) influence the overall dynamics of the network. Scale dependency occurs when network properties change with observational scale. Scale-free networks show a particular frequency distribution of their numbers of nodes and edges, the result being that network properties, such as information flow and network robustness, are the same irrespective of the scale of observation.

Not all nodes are equal, and the particular position of a node in the network can have substantial implications for the links between information and processes at different scales. In a natural resource setting one can think of nodes as agents that experience local environmental variability and edges as representing the sharing of information about their experiences. In computer simulations of agents that manage a resource and share information about their experiences, too much information flow (i.e., very dense networks) can result in a global averaging of management strategies. This leads to efficient management solutions in the short term, but ultimately causes large-scale instead of localized collapses. Too little information flow can result in local sliding-baseline syndromes, as agents may become locked into suboptimal configurations without realizing that better options exist (Bodin and Norberg 2005). In real life situations nodes (people) have different numbers and qualities of social links. Connections to the right or wrong kinds of people (i.e., network position) may substantially influence an individual's capacity to gain useful information and make good decisions. The ability to obtain information from different scales of the system and to develop appropriate cross-scale linkages, together with the capacity to modify external inputs and benefit from them at local scales, provides a strong basis for sustainable development (as shown by Hahn et al. in chapter 4).

Another application of network theory for which scale is a fundamental issue is in the design of networks of protected areas. In recent years there has been a lot of interest in reserve network design (e.g., Possingham, Ball, and Andelman 2000; Meir, Andelman, and Possingham 2004). The scale at which protected areas or "safe havens" are provided must match the dispersal capabilities of the organisms they are expected to protect. Another less obvious component of the problem of reserve network design is understanding the temporal dynamics of landscape change. If nodes in the network blink on or off, as in Conway's game of life (Gardner 1970), the rate at which they do so will influence the number of nodes the network must contain to be viable. Some dynamics (such as more frequent disturbances or a longer regeneration time) will require more nodes to be present if a situation is to be avoided in which the connectivity of the entire network is broken. As changes occur in the areas outside reserves, they influence the likelihood of regional disturbances. Global processes such as climate change and nitrogen deposition occur at a broader scale than reserve networks. Although restricting tenure rights in protected areas can control local processes, such as deforestation or erosion, reserves are as vulnerable as the rest of the landscape to regional changes. For this reason reserve networks cannot ignore changes at the regional scale and expect to avoid them or be buffered against them unless careful consideration has gone into the location and number of nodes and edges in the network. For example, increased connectivity of networks is often perceived as "good," but emerging infectious diseases such as TB or influenza may travel easily through a well-connected network. Ultimately, since the influences that can travel through networks may be either "good" or "bad" from a human perspective, networks with intermediate levels of connectivity may be most resilient—although maintaining such a reserve network will require an additional level of redundancy that may be costly to society.

INFORMATION PROCESSING AND SCALE

Congruence between the scales of internal and external patterns and processes is necessary for effective information processing, as discussed in chapter 5. Organisms have evolved to obtain information about their environment at scales that are relevant to their own survival. In the case of humans, that makes us good at paying attention to our immediate environment (for example, keeping your neighbor's dog out of your yard) but poor at paying attention to potentially more threatening processes on other scales, such as climate change or deforestation. The size of the CAS typically determines the scale of processes that are relevant to it as well as its ability to respond to external influences at an appropriate speed. Systems that are large or very distributed may be harder to perturb, but can also have slower response times. Scale can thus affect the rate of input of information to a system, the rate at which information is processed, and the time until consensus on a single response is reached.

Social and ecological systems have many interacting components that respond at different time scales. The environment that constrains the dynamics of these systems is also made up of multiple variables that change at different spatial and temporal scales. Understanding how the dynamics of slow variables affect systems with faster dynamics has proven to be a major challenge for managers (Carpenter, Ludwig, and Brock 1999), as in the accumulation of phosphate in soils, the reduction of biodiversity, or the erosion of an aspect of social capital such as trust. Systems that have the potential for regime shifts once some threshold is crossed (Walker and Meyers 2004) are of particular concern. Identification of potential thresholds has typically been undertaken in one of two ways: either through exhaustive research that provides the information that is necessary to predict the threshold from first principles, or through analysis of similar case studies in which the threshold was crossed. Chapter 6 presents a third, novel method: Brock, Carpenter, and Scheffer show that information about the temporal variation in a fast variable can be processed to obtain an indication of an approaching threshold. As a system nears the threshold, the amplitude of variation in the fast variable (for example, water clarity) increases and shifts toward slower frequencies. This signal occurs before the threshold is crossed. For many systems, although probably not for all, the signal could be detected early enough to make preventive actions feasible. Chapter 6 offers a new and elegant approach to the problem of predicting the consequences of interactions of variables that operate at different scales.

Another fundamental problem of information processing in a resource management setting is that the complexity of the social-ecological system always exceeds the human capacity for gathering and evaluating information. Thus, decisions must inevitably be made using a scaled-down representation of the real system (chapter 5, Anderies and Norberg). Furthermore, knowledge gathered by people interacting with the system is often based on unique and shared experiences and thus is usually local in nature and limited to a relatively short time period. The limits on human knowledge make it difficult to develop solutions to emerging problems, since conditions may now be so different from the past that previously successful approaches no longer work. Humans typically interpret information based on a contextual understanding. Such interpretations may be unreliable and often change too slowly or too fast. For example, surveys show that fishermen on the east coast of the U.S. consider a sailfish that is a quarter of the size of an average fish from fifty years ago to be a sign of a fishery that is in good health. Mechanisms that ensure that knowledge is passed on over large time scales help to create a more reliable context and can be crucial in preventing such "sliding baseline" syndromes. The same principle holds for rare events that occur at time scales of more than one generation. For instance, Berkes (1998) retells the story of a severe decline in caribou, a traditional game animal for Canadian Chisabi Cree Indians. In that case, the elders drew on their knowledge of a similar situation that had arisen seventy years previously and led the community to make a successful response based on these memories.

The accumulation of knowledge over larger areas and longer time periods can both facilitate and stifle innovation. People may become locked in ways of managing resources that gradually become less and less suited to local conditions. Conversely, when new habits are formed, some behaviors are reinforced and others are lost. The loss of response diversity can reduce the capacity of a society to respond to changing conditions. In such instances the scales at which information is transferred can be important for success, with external information acting as a source of innovation by offering other solutions or even other values that can be communicated as visions.

AN IMPORTANT APPLICATION: SCALE MISMATCHES

Congruence between scales becomes a topic of particular significance in the context of the conservation and management of natural resources. Management generally involves the manipulation and use of a complex, multiscale ecosystem by humans who are themselves members of an organization, family, or other kind of social group. Effective interaction between the social and ecological components of such linked system is contingent on a set of scale-related exchanges between and within the broader system. For example, management interventions are limited by budget and available manpower; monitoring programs take time to yield useful information; resources such as fishes, trees, or grazing may be used or harvested; societal beliefs and needs may drive management priorities; and the flow of information within an organization may occur at many different rates and over many different spatial and temporal scales.

The importance of connections between social and ecological systems becomes particularly obvious when scale mismatches arise. Scale mismatches occur when the scale of environmental variation and the scale of social organization where responsibility for management resides are aligned in such a way that one or more functions of the social-ecological system are disrupted, inefficiencies occur, and/or important components of the system are lost (Cumming, Cumming, and Redman 2006). A fundamental problem for managers is that of matching scales of monitoring and observation to the actions imposed on the system. If decisions are based on different spatiotemporal scales than those the implemented actions will actually affect, the potential for scale mismatch is evident. For example, reductions in nutrient leakage from land uses based on water nutrient content or water clarity may be inadequate if the system has already become locked into a high-nutrient state (Carpenter, Ludwig, and Brock 1999). Similarly, an extractive forest may be viable when the only users are individual members of a nearby village, because the number of trees that is removed is small enough and occurs at a slow enough rate to allow regeneration and observation of the feedbacks that harvesting has on the forest system. If the scale of extraction changes, for instance, by the formation of a logging company with more efficient tools that makes decisions based on regional wood abundances and much more manpower, a scale mismatch can arise in which tree cutting occurs too fast and over too broad an area for effective regeneration to occur. In this example the extractive capacity of the organization increases to a level where it outstrips the provisioning capacity of the ecosystem. Similar problems arise in pastoral systems when cattle herds are large and property sizes are small. Where resolution of a scale mismatch requires a level of mutual cooperation and individual sacrifice, and it can be difficult for institutions to arise at the appropriate scale to manage the use of ecosystem goods and services.

Scale mismatches can also arise in ecosystems without human intervention, although they are typically short-lived (in relative terms) because of the adaptive nature of the system and the relatively tight feedbacks between system components. For example, gradual increases in populations of large, long-lived herbivores and carnivores, such as elephants and killer whales, can result in a temporal mismatch between the resource requirements of large organisms and the production of forage and prey items at finer scales. This can lead to increasing resource degradation, ultimately causing a collapse of populations of the larger animals (Cumming et al. 1997). Such a collapse realigns ecological scales and restores system function.

Scale mismatches assume particular importance where positive (amplifying) or negative (regulating) feedbacks between processes at different scales are central to system organization and function. Scale mismatches can result in the disconnection of feedbacks or in feedbacks changing from tight to loose or vice versa. Changes in feedbacks can have profound consequences for local resource use and management. An obvious example arises when a management agency has too limited a mandate to carry out actions that are clearly necessary over a larger area, such as prescribed burning or control of an invasive species. In this instance the scale mismatch can provide a strong incentive for the development of broader-scale, more inclusive institutions and, indeed, has been one of the driving forces behind the development of community-based conservation.

GENERAL PRINCIPLES AND QUESTIONS OF SCALE IN CASs

To summarize, this book contains a number of important messages about scale and its relevance for people who have an interest in understanding and managing complex systems. Some of these insights are novel and others are widely accepted. In the widely accepted category are the observations that both our perceptions of patterns and their underlying nature are likely to change with scale (with the rare exception of truly fractal patterns), that the magnitudes, rates, and directions of processes and relationships may change with scales, and that the scales that are generally most relevant to understanding a given problem are the scales that are closest to the scale of the problem itself. Scientists have for many decades looked to finer scales to explain mechanisms and to broader scales to understand limits and constraints on systems that range from ecosystems to individual cells. Consideration of scale-related asymmetries and differences in scaling functions have offered many useful insights into limitations on processes, for example, in helping us to understand why big animals don't get bigger (Peters 1983) and why big, fierce animals are rare (Colinvaux 1979).

Theories about emergent properties and cross-scale feedbacks are less widely accepted than the more straightforward applications of scaling theory, although these exciting new developments are gaining an increasing number of adherents. Some of the more interesting generalities to emerge are the ideas that the dynamics of complex adaptive systems can be interpreted through understanding changes in feedback loops within and between scales (Levin 1999), that the fundamental controls on system behavior, those that cause shifts in system state, are often variables that change quite slowly, such as phosphorus levels in lakes or soil fertility in agricultural systems (Scheffer et al. 2001), and that diversity plays a pivotal role in system resilience and sustainability at different scales during times of perturbation (Levin 1999; Gunderson and Holling 2002).

In the case of management and other applications of scaling theories, the value of taking scale-related issues into account is very clear (Carpenter 2002). Most managers are familiar with the specter of looming ecological problems they have neither the resources nor the mandate to manage (e.g., Nepstad et al. 1999). Such problems include such diverse issues as global climate change, control of expanding elephant populations, and eradicating fire ants. Scale mismatches typically lead to substantial degradation of important resources. Identifying the central problem as a scale-related issue can lead more quickly to an effective management strategy, although creating the kinds of adapting, learning institutions that are most effective in the management of complex systems is often the hardest part of the problem.

The study of scale and scaling in complex adaptive systems remains an exciting frontier that presents many challenges. One of the largest of these is the derivation of general principles and predictive relationships. Many of the scaling relationships that have been described in landscapes are unique, lacking in generality across space and falling apart during times of system-wide change. Another important challenge for scaling theory will be to link changes in social and ecological systems at different scales to differences in both environmental configuration and environmental composition (Flather and Bevers 2002). We need to advance scaling theory to the point where it can cope with system reorganization and regeneration as well as steady growth, which has been its main focus to date. As Holling, Gunderson, and Peterson (2002) suggest, overall system dynamics may depend not only on the scales and magnitudes of different processes but also on their relative alignment to one another in terms of resources and connectivity. Certain kinds of system change may only occur when there is instability at several different scales, allowing a radical transformation of the system in the absence of stabilizing feedbacks. If system dynamics at each hierarchical level are viewed as waves, scaling studies need to understand not only what drives variability at each scale but also what leads to synchrony (or lack of it) between peaks and troughs. For example, ground cover and canopy cover in a woodland will not necessarily be in synchrony; fires that are hot enough to reorganize the system by killing woody species may require both high-fuel ground cover and high-fuel canopy cover.

Understanding feedbacks and compartmentalization is one of the areas we consider promising in research into scale and CASs. As system components interact with one another across scales, groups of units tend to show some level of spatial and/or temporal synchrony, often based on their common membership in a subcategory of a hierarchy. For example, flowering plants and their insect pollinators may group into functional compartments that are defined by the presence of strong interactions relative to other species in the same landscape (Dicks, Corbet, and Pywell 2002; Prado and Lewinsohn 2004). Compartmentalization is also important in social and economic systems; for example, although the different states in the U.S. and Canada are autonomous in some ways, their collective fortunes rise and fall with the national economy. Such compartmentalization makes the economies of Florida and Texas different from those of Ontario and Alberta.

One of the oldest debates in ecology is centered on the question of whether most ecological communities should be viewed as compartments that have some degree of autonomy from the ecosystem and the landscape (Clements 1936) or as assemblages of individual members of different species (Gleason 1926). The perspective that is adopted in any given circumstance has some important practical implications. In managing for the maintenance of ecosystem services, we often lose local services long before we lose regional or global services because ecosystem services are provided by local populations. If communities are semi-autonomous, local extinctions may be as harmful at the local scale as global extinctions. By contrast, if communities are merely species assemblages, then restoration and reintroduction activities stand a much higher chance of success. Understanding the interactions and feedback loops that occur within hierarchies is thus one of the advances that seems necessary for us to be able to develop a predictive understanding of a variety of scale-related phenomena, including scale breaks, thresholds, and nonlinearities within systems of interest.

REFERENCES

Allen, C. R., E. A. Forys, and C. S. Holling. 1999. "Body Mass Patterns Predict Invasions and Extinctions in Transforming Landscapes." *Ecosystems* 2:114–121.

Allen, T. F. H., and T. B. Starr. 1982. *Hierarchy: Perspectives for Ecological Complexity.* Chicago: University of Chicago Press.

Andersson, C., S. Rasmussen, and R. White. 2002. "Urban Settlement Transitions." *Environment and Planning B: Planning and Design* 29:841–865.

Bellehumeur, C., and P. Legendre. 1998. "Multiscale Sources of Variation in Ecological Variables: Modeling Spatial Dispersion, Elaborating Sampling Designs." *Landscape Ecology* 13:15–25.

Berkes, F. 1998. "Indigenous Knowledge snd Resource Management Systems in the Canadian Subarctic." In F. Berkes and C. Folke, eds., *Linking Social and Ecological Systems: Management Practices and Social Mechanisms for Building Resilience*, pp. 98–128. Cambridge: Cambridge University Press.

Berry, B. 1961. "City Size Distributions and Economic Development." *Economic Development and Cultural Change* 9:573–588.

Bodin, Ö., and J. Norberg. 2005. "Information Network Topologies for Enhanced Local Adaptive Management." *Environmental Management* 35:175–193.

Borcard, D., P. Legendre, C. Avois-Jacquet, and H. Tuomisto. 2004. "Dissecting the Spatial Structure of Ecological Data at Multiple Scales." *Ecology* 85:1826–1832.

Brown, J. H. 1995. *Macroecology.* Chicago: University of Chicago Press.

Carpenter, S. R. 2002. "Ecological Futures: Building an Ecology of the Long Now." *Ecology* 83:2069–2083.

Carpenter, S. R., D. Ludwig, and W. A. Brock. 1999. "Management of Eutrophication for Lakes Subject to Potentially Irreversible Change." *Ecological Applications* 9:751–771.

Chave, J., and S. Levin. 2003. "Scale and Scaling in Ecological and Economic Systems." *Environmental and Resource Economics* 26:527–557.

Chave, J., H. C. Muller-Landau, and S. A. Levin. 2002. "Comparing Classical Community Models: Theoretical Consequences for Patterns of Diversity." *American Naturalist* 159:1–23.

Clements, F. E. 1936. Nature and Structure of the Climax. *Journal of Ecology* 24:252–284.

Colinvaux, P. A. 1979. "Why Big Fierce Animals Are Rare: An Ecologist's Perspective." Princeton: Princeton University Press.

Connell, J. H. 1978. "Diversity in Tropical Rain Forests and Coral Reefs." *Science* 199:1302–1310.

Cumming, D. H. M., and G. S. Cumming. 2003. "Ungulate Community Structure and Ecological Processes: Body Size, Hoof Area, and Trampling in African Savannas." *Oecologia* 134:560–568.

Cumming, D. H. M., M. B. Fenton, I. L. Rautenbach, R. D. Taylor, G. S. Cumming, M. S. Cumming, J. M. Dunlop, A. G. Ford, M. D. Hovorka, D. S. Johnston, M. Kalcounis, Z. Mahlangu, and C. V. R. Portfors. 1997. "Elephants, Woodlands, and Biodiversity in Southern Africa." *South African Journal of Science* 93:231–236.

Cumming, G. S., and J. Collier. 2005. "Change and Identity in Complex Systems." *Ecology and Society.* http://www.ecologyandsociety.org/vol10/iss1/art29/.

Cumming, G. S., D. H. M. Cumming, and C. L. Redman. 2006. "Scale Mismatches in Social-Ecological Systems: Causes, Consequences, and Solutions." *Ecology and Society.* http://www.ecologyandsociety.org/vol11/iss1/art14/.

Cumming, G. S., and T. D. Havlicek. 2002. "Evolution, Ecology, and Multimodal Distributions of Body Size." *Ecosystems* 5:705–711.

Dicks, L. V., S. A. Corbet, and R. F. Pywell. 2002. "Compartmentalization in Plant-Insect Flower Visitor Webs." *Journal of Animal Ecology* 71:32–43.

Dungan, J. L., J. N. Perry, M. R. T. Dale, P. Legendre, S. Citron-Pousty, M.-J. Fortin, A. Jakomulska, M. Miriti, and M. S. Rosenberg. 2002. "A Balanced View of Scale in Spatial Statistical Analysis." *Ecography* 25:626–640.

Espinosa-Fuentes, M. L., and C. Flores-Coto. 2004. "Cross-shelf and Vertical Structure of Ichthyoplankton Assemblages in Continental Shelf Waters of the Southern Gulf of Mexico." *Estuarine Coastal and Shelf Science* 59:333–352.

Flather, C. H., and M. Bevers. 2002. "Patchy Reaction-Diffusion and Population Abundance: The Relative Importance of Habitat Amount and Arrangement." *American Naturalist* 159:40–56.

Gardner, M. 1970. "Mathematical Games: The Fantastic Combinations of John Conway's New Solitaire Game 'Life.'" *Scientific American* 223:120–123.

Gaston, K. J. 2000. "Global Patterns in Biodiversity." *Nature* 405:220–227.

Gibson, C. C., E. Ostrom, and T. Ahn. 2000. "The Concept of Scale and Human Dimensions of Global Change: A Survey." *Ecological Economics* 32. 217–239.

Gleason, H. A. 1922. On the Relation Between Species and Area." *Ecology* 3:158–162.

Gleason, H. A. 1926. "The Individualistic Concept of the Plant Association." *Bulletin of the Torrey Botanical Club* 53:7–26.

Gunderson, L., and C. Holling, eds., 2002. *Panarchy: Understanding Transformations in Human and Natural Systems.* Washington, DC: Island.

Haskell, J. P., M. Ritchie, and H. Olff. 2002. "Fractal Geometry Predicts Varying Body Size Scaling Relationships for Mammal and Bird Home Ranges." *Nature* 418:527–530.

Havlicek, T. D., and S. R. Carpenter. 2001. Pelagic Size Distributions in Lakes: Are They Discontinuous? *Limnology and Oceanography* 46:1021–1033.

Holling, C. S. 1973. "Resilience and Stability of Ecological Systems." *Annual Review of Ecology and Systematics* 4:1–23.

Holling, C. S. 1992. "Cross-scale Morphology, Geometry, and Dynamics of Ecosystems." *Ecological Monographs* 62:447–502.

Holling, C. S. 2001. "Understanding the Complexity of Economic, Ecological, and Social Systems." *Ecosystems* 4:390–405.

Holling, C. S., L. H. Gunderson, and G. D. Peterson. 2002. "Sustainability and Panarchies." In L. H. Gunderson and C. S. Holling, eds., *Panarchy: Understanding Transformations in Human and Natural Systems*, pp. 63–102. Washington, DC: Island.

Hubbell, S. P. 2001. *The Unified Neutral Theory of Biodiversity and Biogeography.* Princeton: Princeton University Press.

Hurlbert, S. H. 1984. "Pseudoreplication and the Design of Ecological Field Experiments." *Ecological Monographs* 54:187–211.

Huston, M. 1994. *Biological Diversity: The Coexistence of Species on Changing Landscapes.* Cambridge: Cambridge University Press.

Huston, M. 1997. "Hidden Treatments in Ecological Experiments: Re-evaluating the Ecosystem Function of Biodiversity." *Oecologia* 108:345–350.

Isaaks, E. H., and R. M. Srivastava. 1989. *An Introduction to Applied Geostatistics.* Oxford: Oxford University Press.

Koestler, A. 1967. *The Ghost in the Machine.* New York: MacMillan.

Kondoh, M. 2001. "Unifying the Relationships of Species Richness to Productivity and Disturbance." *Proceedings of the Royal Society, Biological Sciences* 268:269–271.

LaBarbera, M. 1989. "Analyzing Body Size as a Factor in Ecology and Evolution." *Annual Review of Ecology and Systematics* 20:97–119.

Lawton, J. H. 1999. "Are There General Laws in Ecology?" *Oikos* 84:177–192.

Legendre, P., and L. Legendre. 1998. *Numerical Ecology.* 2d ed. London: Elsevier.

Levin, S. A. 1992. "The Problem of Pattern and Scale in Ecology." *Ecology* 73:1943–1967.

Levin, S. A. 1999. "Fragile Dominion: Complexity and the Commons." Cambridge: Perseus.

Longley, P. A., M. Batty, and J. Shepherd. 1991. "The Size, Shape and Dimensions of Urban Settlements." *Transactions of the Institute of British Geographers* 16:75–94.

Mandelbrot, B. B. 1983. *The Fractal Geometry of Nature.* New York: Freeman.

Marquet, P. A., and H. Cofre. 1999. "Large Spatial and Temporal Scales in the Structure of Mammalian Assemblages in South America: A Macroecological Approach." *Oikos* 85:299–309.

Mbow, C., T. T. Nielsen, and K. Rasmussen. 2000. "Savanna Fires in East-Central Senegal: Distribution Patterns, Resource Management and Perceptions." *Human Ecology* 28:561–583.

Meir, E., S. Andelman, and H. Possingham. 2004. "Does Conservation Planning Matter in a Dynamic and Uncertain World?" *Ecology Letters* 7:615–622.

Meredith, J. R. 2003. "Sprawl and the New Urbanist Solution." *Virginia Law Review* 89:447–503.

Nepstad, D. C., A. Verissimo, A. Alencar, C. Nobre, C. Lima, P. Lefebvre, W. Schlesinger, C. Potter, P. Mourinho, E. Mendoza, M. Cochrane, and V. Brooks. 1999. "Large-scale Impoverishment of Amazonian Forests by Logging and Fire." *Nature* 398:505–508.

O'Neill, R. V., D. DeAngelis, J. Waide, and T. F. H. Allen. 1986. A *Hierarchical Concept of Ecosystems.* Princeton: Princeton University Press.

Peters, R. H. 1983. *The Ecological Implications of Body Size.* New York: Cambridge University Press.

Peterson, G. D., C. R. Allen, and C. S. Holling. 1998. "Ecological Resilience, Biodiversity, and Scale." *Ecosystems* 1:6–18.

Pianka, E. R. 1966. "Latitudinal Gradients in Species Diversity: A Review of Concepts." *American Naturalist* 910:33–46.

Pickett, S. T. A., and M. L. Cadenasso. 2002. "The Ecosystem as a Multidimensional Concept: Meaning, Model, and Metaphor." *Ecosystems* 5:1–10.

Polis, G. A., and D. R. Strong. 1996. "Food Web Complexity and Community Dynamics." *American Naturalist* 147:813–846.

Polis, G. A., M. E. Power, and G. R. Huxel. 2004. *Food Webs at the Landscape Scale.* Chicago: University of Chicago Press.

Possingham, H., I. Ball, and S. Andelman. 2000. "Mathematical Methods for Identifying Representative Reserve Networks." In S. Ferson and M. Burgman, eds., *Quantitative Methods for Conservation Biology,* pp. 291–305. New York: Springer.

Prado, P. I., and T. M. Lewinsohn. 2004. "Compartments in Insect Plant Associations and Their Consequences for Community Structure." *Journal of Animal Ecology* 73:1168–1178.

Ritchie, M. 1998. Scale-Dependent Foraging and Patch Choice in Fractal Environments." *Evolutionary Ecology* 12:309–330.

Ritchie, M., and H. Olff. 1999. "Spatial Scaling Laws Yield a Synthetic Theory of Biodiversity." *Nature* 400:557–560.

Saetre, P. 1999. "Spatial Patterns of Ground Vegetation, Soil Microbial Biomass and Activity in a Mixed Spruce-Birch Stand." *Ecography* 22:183–192.

Salmaso, N. 1996. "Seasonal Variation in the Composition and Rate of Change of the Phytoplankton Community in a Deep Subalpine Lake (Lake Garda, Northern Italy): An Application of Nonmetric Multidimensional Scaling and Cluster Analysis." *Hydrobiologia* 337:49–68.

Salthe, S. 1985. *Evolving Hierarchical Systems: Their Structure and Representation.* New York: Columbia University Press.

Scheffer, M., S. R. Carpenter, J. A. Foley, C. Folke, and B. Walker. 2001. "Catastrophic Shifts in Ecosystems." *Nature* 413:591–596.

Scheiner, S., S. B. Cox, M. Willig, G. Mittelbach, C. Osenberg, and M. Kaspari. 2000. Species Richness, Species-Area Curves and Simpson's Paradox." *Evolutionary Ecology Research* 2:791–802.

Shumway, R. H., and D. S. Stoffer. 2000. *Time Series Analysis and Its Applications.* New York: Springer.

Simon, H. A. 1962. "The Architecture of Complexity." *Proceedings of the American Philosophical Society* 106:467–482.

Stauffer, D. 1985. *Introduction to Percolation Theory.* London: Taylor and Francis.

Stommel, H. 1963. "The Varieties of Oceanographic Experience." *Science* 139: 572–576.

Tarboton, D. G., R. L. Bras, and I. Rodriguez-Iturbe. 1988. "The Fractal Nature of River Networks." *Water Resources Research* 24:1317–1322.

Turner, M. G., R. H. Gardner, and R. V. O'Neill. 2001. *Landscape Ecology in Theory and Practice: Pattern and Process.* New York: Springer.

Waide, R. B., M. R. Willig, C. F. Steiner, G. Mittelbach, L. Gough, S. I. Dodson, G. P. Juday, and R. Parmenter. 1999. "The Relationship Between Productivity and Species Richness." *Annual Review of Ecology* 30:257–300.

Walker, B., and J. Meyers. 2004. Thresholds in Ecological and Social Ecological Systems: A Developing Database." *Ecology and Society* 9:3.

West, G. B., J. H. Brown, and B. J. Enquist. 1997. "A General Model for the Origin of Allometric Scaling Laws in Biology." *Science* 276:122–126.

Westley, F., S. R. Carpenter, W. A. Brock, C. S. Holling, and L. H. Gunderson. 2002. "Why Systems of People and Nature Are Not Just Social and Ecological Systems." In L. H. Gunderson and C. S. Holling, eds., *Panarchy: Understanding Transformations in Human and Natural Systems*, pp. 103–119. Washington, DC: Island.

White, R., and G. Engelen. 1993. "Cellular Automata and Fractal Urban Form: A Cellular Modelling Approach to the Evolution of Land-Use Patterns." *Environment and Planning A* 25:1175–1199.

Wiens, J. A. 1989. "Spatial Scaling in Ecology." *Functional Ecology* 3:385–397.

With, K. A., R. H. Gardner, and M. G. Turner. 1997. "Landscape Connectivity and Population Distributions in Heterogeneous Environments. *Oikos* 78:151–169.

Wu, J. 1999. "Hierarchy and Scaling: Extrapolating Information Along a Scaling Ladder." *Canadian Journal of Remote Sensing* 25:367–380.

Wu, J., and O. L. Loucks. 1995. "From Balance of Nature to Hierarchical Patch Dynamics: A Paradigm Shift in Ecology." *Quarterly Review of Biology* 70:439–466.

Wu, J., D. E. Jelinski, M. Luck, and P. Tueller. 2000. "Multiscale Analysis of Landscape Heterogeneity: Scale Variance and Pattern Metrics." *Geographic Information Systems* 6:6–19.

COMPLEXITY THEORY FOR A SUSTAINABLE FUTURE
CONCLUSIONS AND OUTLOOK

Jon Norberg and Graeme S. Cumming

THE NATURAL WORLD is characterized by diversity and interactions, continuous change, learning, and the unexpected. In dealing with the complexities of living systems we face the task of "exploring the known, the unknown and the unknowable" (Levin 2002). In contrast to previous conceptual theories or worldviews (such as "nature in balance") that emphasize predictable and nonchanging end-states that are temporarily offset by exogenous disturbances, complexity theory has been developed with the concepts of change, surprise, and adaptation at its core. When applying these concepts to problems of sustainability, they must be used in a way that is integrative and transdisciplinary (Clark and Dickson 2003). They also need to be communicative in the sense that they foster exchange and highlight differences in the ways that disparate disciplines and participants have learned to deal with the characteristics of complex systems. Rather than being concerned with the optimization of single quantities, the focus of applied complexity theory is on building the capacity to act proactively and respond to change effectively (Kates and Clark 1996; Berkes, Colding, and Folke 2003; Folke et al. 2005).

While much remains to be learned about the natural systems that support human life, it has become abundantly clear that failures to sustainably manage natural resources often occur as a consequence of a lack of appreciation and understanding of relevant social processes, rather than insufficient knowledge of underlying natural processes (Folke et al. 2005). Simple optimization-oriented scientific solutions, such as the maximum sustainable yield concept, often fail because they ignore both the complexity of the ecosystem functions that support the target species and the social dimensions that dictate how people organize around top-down institutions (Ludwig,

Hilborn et al. 1993; Dietz, Ostrom, and Stern 2003). Living systems continuously change because of selective processes, environmental heterogeneity, and the introduction of novelty. Policy solutions therefore become limited in time and in their effectiveness. Historical fishing has not only reduced fish stocks but has also fundamentally and perhaps irreversibly altered the structure of the food web and the genetic makeup of the species concerned (Jackson et al. 2001). Furthermore, human understanding and perception undergo continuous adaptation and change, often favoring short-term solutions that may exacerbate the fundamental problem (Holling and Meffe 1996) by creating lock-in adaptations to less desirable states (Pauly 1995). If the mistakes of the past are not to repeat themselves in the future, we need to understand these properties of complex adaptive systems and to develop tools for communication between scientists, policy makers, and appropriators. Complexity theory offers some simple guidelines and rules of thumb that could, if widely accepted, provide the basis for a quiet revolution in the theory and practice of the management of natural resources.

Within the field of natural resource management, concepts such as human-environment systems, ecosocial systems, or socioecological systems have been used (Folke, Hahn et al. 2005). Berkes and Folke (1998) suggested that the term *social-ecological system* (SES) would emphasize the integrated concept of humans in nature while not reducing any discipline to a prefix. Social-ecological systems are not social systems plus ecological systems (Westley et al. 2002). A social-ecological view has major implications for the way in which one views system dynamics in comparison to separate social or ecological perspectives. For example, a nonlinear feedback in sediment responses to oxygen can cause a lake to flip into a eutrophied state if phosphate loadings become too high. This is often perceived as a regime shift for the ecological system (Carpenter and Gunderson 2001). From a social-ecological perspective the system has only shifted if, in addition to the change in the lake, the social capacity and willingness to bring the lake back to a noneutrophied state has been changed (Bodin and Norberg 2005). As long as both willingness and manipulative capacity are maintained, the attractor state for the social-ecological system remains the noneutrophied state even though the lake may temporarily become eutrophic. The importance of this insight is that efforts to manage the state of the lake need to focus as much on the social system as the natural system.

The true nature of sustainability—what it means, and how it should be achieved—is only gradually becoming apparent (Costanza 1991; Dasgupta

2001). Sustainability itself is an elusive concept (Goldman 1995) and there has been a tendency in many fields to focus on single aspects of sustainability while ignoring the larger and more complex system to which they belong. For example, attempts to conserve biodiversity are often ecologically sound but perceived as socially and economically inefficient. In our view, sustainability is about taking actions in the present that are intended to maintain and enhance the current and future resilience of desired trajectories of development of the social-ecological system, while meeting a range of more immediate goals that are important for social and ecological persistence at short time scales and over smaller areas. The agenda of sustainability must inevitably incorporate understanding from ecology, social sciences, and economics, but in such a way that disciplinary values and assumptions are subordinate to the overall agenda. For example, the aim of maximizing short-term wealth often guides decisions about land use change, with more sustainable and longer-term perspectives being brushed aside by special interest groups; and the same kind of logic often guides the selection of discounting rates (i.e., the reduction in value of an item as the time until it is received increases) in economic calculations. Humanitarian values such as equity and ethics matter not only for their intrinsic worth but also because they often contribute to long-term sustainability by correcting asymmetries that might ultimately result in system-wide collapse and reorganization.

Complexity theory provides a transdisciplinary, value-neutral body of ideas that together can be used to understand a wide range of macroscopic properties of the seemingly insurmountable complexity of the world in which we live. Complexity theory can not be claimed by any single discipline, since its central tenets have been discovered and developed in parallel disciplines, albeit with differences in terminology and through time (Folke 2006).

This book does not claim to provide answers to the most pressing questions of our time. It does, however, present a number of interesting perspectives, tools, and ideas that together form the beginnings of the theoretical and applied approaches that we believe will ultimately provide those answers. As the chapters in this book make abundantly clear, the field of sustainability research is progressing rapidly but still has many crucial and fascinating questions to answer. In this chapter we discuss the findings of previous chapters in the light of two questions. First, what kinds of general concepts emerge from the chapters in the book, and how do they contribute to complex systems theory and its applications? And, second, which areas of study and application seem to be most in need of further research, and how might they progress in the future?

COMPLEXITY THEORY: GENERAL CONCEPTS

A CAS perspective is useful because it provides an understanding of *general* macroscopic dynamics of complex adaptive systems that can be used to understand *particular* systems by comparative approaches. We have already mentioned the usefulness of thinking about regimes within which systems organize around certain self-reinforcing processes (Carpenter 2003). This reinforcement prevents or reduces the likelihood that another regime (system configuration) will occur. Without perturbations or directional changes in the system's environment, systems will in theory move toward one attractor state within a regime. An attractor state can be dynamic in itself, for example if it fluctuates in response to changing climatic drivers or economic factors such as market prices or exchange rates. When internal feedback processes change, thresholds can be crossed, new feedbacks become dominant, and new regimes can occur in which different conditions drive the interactions among system components. Although it may be difficult to identify feedbacks and thresholds, many systems display these kinds of dynamics (Walker and Meyers 2004). In other cases, systems will be dominated by a single feedback at key spatial and temporal scales and thus will lack the potential for shifts between regimes (Briske, Fuhlendorf, and Smeins 2003).

Regardless of their potential for regime shifts, all living systems experience gradual changes that are driven by factors that are more or less external to the system. Adaptive capacity (which extends the concept of response capacity; Elmqvist et al. 2003) is a measure of how readily important system structures, processes, and components can be sustained under changing conditions. System elements are kept in place by means of adaptive (in its general meaning) responses within system components, groups of components within a system, or whole systems. Adaptive responses include reactive responses, such as succession within biotic communities, or proactive behavior such as taking precautions against future adversity. Adaptive capacity in general can be defined as the outcome of having different options and having the potential to switch between them. Thus, even if options are abundant, a system can lack adaptive capacity if change between options is not possible. Overregulated organizations often fall into this category if decision-making procedures become so slow that change outpaces the response (Arrow, McGrath, and Berdahl 2000). On the other hand, the willingness to change may exist, but suitable options may not. The concept of adaptive capacity is not directly linked to the concept of regime shifts, but it is of great importance in understanding how regime shifts can be avoided and how reorganization occurs after a regime shift.

Consideration of selective processes in SESs emphasizes qualitative change in systems that have the potential for fundamental changes in their dynamics as well as for changes in the larger constraints under which the system operates. Quantitative measures of local variables, such as catch rates in fisheries, do not usually tell us much about global changes like erosion of the capacity of the fish population to reproduce or the spatial dynamics of interacting subpopulations. Selective processes that act on system components can change the fundamental dynamics of the system and particularly its capacity to respond to future changes in conditions. Often the effects of selective processes are relatively slow and difficult to detect. Other examples include the erosion of social capital (Putnam 2004; Misztal 1996) and declines in regional biodiversity (Ehrlich and Ehrlich 1981). Selective processes often respond to the slowly changing attractor states of the system and thus frame the direction in which the system is heading. By contrast, rapid dynamics and turbulent change are more likely to occur during regime shifts, which take place once the slow variables that determine the system's current domain have finally crossed a critical threshold.

The general concepts of resilience, regime shifts, adaptive capacity, and selective processes are useful when thinking about the dynamics of complex adaptive systems. They are used in nearly all of the chapters in this volume and have facilitated the development of many of the overarching conclusions that we discuss next.

CONSEQUENCES OF CONTROLLING SPATIOTEMPORAL VARIABILITY

For many thousands of years, humans have stayed alive by reducing spatial and temporal uncertainties in basic needs. For example, most modern societies have created a relatively stable supply of food and water through the use of fossil fuels, trade systems, and information as well as technological innovations. In creating a more predictable local environment through regional linkages, societies have been able to improve efficiency by concentrating efforts toward locally efficient production systems and specialized livelihoods without facing substantial risks if they cannot be sustained when conditions change. Relying on more efficient but fewer sources of income, however, increases sensitivity to changes as systems become more dependent on critical components. If the expected variability in conditions is known and the society has invested in capacity to deal with it, however, this strategy can be (and has been) very successful.

Control of local uncertainty in living conditions has had the unfortunate side effect of increasing the spatial scales of both dependence and influence of an increasing human population (Holling and Meffe 1996; Vitousek et al. 1997). The environment today is influenced by humans on a relatively large scale, such that the pattern and magnitude of variability in environmental and social drivers has also changed (Jackson et al. 2001; Diamond 2005; MEA 2005). A growing body of theory suggests that for simple linear systems controlling variability at one scale may induce higher amplitudes at other frequencies (Skogestad and Postlethwaite 1996). This outcome is sometimes called the conservation law of fragility or highly optimized tolerance (Carlson and Doyle 1999; Robert, Carlson, and Doyle 2001). Of particular interest for complex systems such as social-ecological systems (SES) is that many of these conclusions seem to hold for nonlinear systems as well (Brock and Durlauf 2005). In fact, controlling variability in both space and time seems to induce higher amplitudes in fluctuations at larger temporal scales (Brock and Xepapdeas 2004; Carpenter and Brock 2006). Because of the inherent dynamics of complex adaptive systems (Holling 1992), controlling for spatiotemporal conformity at one scale (typically, favoring a system state with efficient, high-volume and low-disturbance production) creates regional sensitivity and strong positive feedbacks by which small disturbances are amplified and easily spread. An ecological example of this phenomenon is that of forests in which local fires are actively mitigated, an action that may lead to catastrophic and regionwide wildfires once biomass has accumulated at a large scale. The combination of increased sensitivity of local systems toward change and increased variability in external drivers at larger scales suggests that social-ecological systems in general may face greater risks of crossing internal thresholds than ever before.

ANTICIPATING CHANGE

Now that anthropogenic actions have global effects, the possibility of learning from our mistakes is being lost; one global mistake may be catastrophic. This increasing risk leaves us more dependent than ever on our capability for proactive behavior. The heart of proactive behavior lies in an ability to anticipate change and to understand the processes that govern system dynamics. The willingness of society to act on perceived changes in the future is limited, however, especially if there are large and immediate economic interests involved and the causal relationships driving change are complex. For simple relationships and very specific economic interests, as in the resolution of chlorofluorocarbon (CFC) production by the Montreal Protocol, social responses can often be mobilized in time (even though the response in the CFC case was more reactive than proactive). Complexity theory suggests that the key to the capacity of society to act in time lies in 1. learning to identify the right signals and embed the appropriate response capacity in efficient institutional frameworks; and 2. nurturing the capacity to adapt to change and transformations. Much recent research has emphasized the relevance of historical time records, for example, in relation to climate change, in analyzing long-term trends and changes in variability. A fundamental problem, however, is the question of how to predict an impending change in fundamental feedback processes (i.e., a regime shift). Chapter 6, by Brock, Carpenter, and Scheffer, provides a novel and promising approach that utilizes a general phenomenon in complex systems (i.e., an increasing amplification of variability in external driving variables as the system approaches a threshold between two regimes of attraction) to predict when a system is approaching a threshold. While this approach still needs further testing and using historical time series, it may provide a fundamentally new way of predicting impending transformations in complex systems.

NURTURING SOURCES OF ADAPTIVE CAPACITY

A common problem in understanding the role of diversity in system performance is the narrow focus on *current* diversity as opposed to a focus on the processes that *maintain and diminish* diversity. Conserving current diversity (i.e., a particular set of species or customs) in a situation where conditions change is bound to be maladaptive in the long run since a different set of species or rules might cope better with new conditions (Norberg et al.; chapter 2). In order to understand the functional role of diversity in SESs, it is imperative to focus on how diversity is maintained and what are the selective processes that act on it. Processes that diminish diversity are often perceived as negative, yet they provide the "fuel" for adaptation as they open up opportunities for change. Conservation strategies based on the "balance of nature" view will often focus on avoiding change, even if (for example) a species is on the edge of its range and the range is slowly shifting in response to climatic change (Davis and Shaw 2001). Nurturing adaptive capacity in systems in general thus needs to recognize the counteracting roles of processes that respectively increase or decrease diversity. Increases in diversity provide variety that enhances the potential for adaptation to changing conditions, while selection ensures that change can take place by removing old system components or interactions.

Sustaining diversity is closely related to the idea of novelty. Novelty is, as discussed by Anderies and Norberg (chapter 5), a relative issue and is dependent on the perspective of the observer. In the management of natural resources, the spatial focus typically lies at local or regional scales and thus new species can emerge by local evolutionary processes (mutation and recombination) or by immigration from an external species pool. Institutional diversity can be sustained by transferring knowledge from external sources, applying current local knowledge in a different context, or developing "true novelty" in the sense that a new solution is conceived by innovation. Thus, one can distinguish between sources of diversity that are 1. based on the local already available pool (i.e., recombination and mutation in species or context transfer in institutions) and 2. independent of what is locally available. The first will provide solutions that are different yet similar and that would be most useful during gradual and adaptive change. The second is more likely to provide radically new designs, something that is of particular value during turbulent and abrupt change.

Adaptive capacity is needed to respond to gradual change, but situations also exist where the present diversity does not provide the variety of alternatives needed. This makes the system sensitive, as one single alternative (a species or institution) can become dominant simply because of a lack of competitive alternatives rather than by being particularly effective as a response to new conditions. Governance in SESs that attempts to nurture adaptive capacity thus needs to recognize that 1. change and disturbance are fundamental for sustaining adaptive capacity, 2. adaptive capacity is maintained by the interplay between processes that sustain and diminish diversity, and 3. different sources of diversity may be important under different disturbance regimes.

GOVERNING SELF-ORGANIZATION: PROVIDING VERSUS ENABLING SOLUTIONS

Social-ecological systems are to a large degree self-organizing systems, and attempts to control them will result in responses, for example, by inducing cheating behavior as a result of eroded trust in top-down or imposed institutions (Dietz, Ostrom, and Stern 2003). A complexity perspective on governance in social-ecological systems must therefore be aware of the processes that govern self-organization in SES. The fundamental change in perspective that a focus on self-organization leads to is that solutions must be enabled and not provided. There are numerous examples in this volume of how the enabling of solutions may be achieved. Chapter 7 (Lebel and Bennett) elaborately discusses the opportunities and fallacies of the participatory approach; Hahn et al. (chapter 4) highlight the role of stewardship and local networks for creating the social capital that enables local solutions to complex problems. These issues have also been highted in earlier papers by Westley (2002) and Westley and Miller (2003). In her latest book Lin Ostrom (2005) elaborates on the conditions needed for enabling collective action that has the potential for crafting sustainable governance solutions.

FUTURE RESEARCH CHALLENGES

Complexity theory and its applications constitute a rapidly growing area of research with many unexplored areas and many exciting challenges. Throughout this book, we and other contributing authors have focused on themes that are currently of interest and potential application in the context of social-ecological sustainability and resilience. In closing, we focus on a few emerging areas of research and conceptual development that we think will be of particular importance for studies of SES in the coming years.

It is perhaps too early to talk of a paradigm shift in conservation and related disciplines. However, one of the major challenges ahead is to transform conservation biology from its roots as an applied discipline to one that is informed by theory and capable of generalization. Conservation has historically been dominated by local and species-level concerns, a lack of hypothesis development and testing, and innumerable case studies: while its focus on achieving tangible results has been valuable, it has all too often wandered down blind-ending pathways as a result of a weak or nonexistent theoretical basis. The chapters contained in this book demonstrate that conservation both can and should be informed by theory and that, ultimately, the theory required for conservation must be deeper and more complex than ecological theory, because conservation will inevitably include relevant aspects of disciplines such as economics and sociology in addition to ecological understanding.

Many of the research challenges raised by this book revolve around complex system dynamics, self-organization, the causes and consequences of spatiotemporal variation in linked social-ecological systems, and social-ecological linkages in relation to governance. In each case one of the largest gaps in our understanding concerns the interactions between social and ecological components of the system and the ways in which feedbacks and cross-scale effects result in complex behaviors. Governance cannot be viewed as external to the SES; institutions and organizations arise from within the SES and must self-regulate while remaining within the limits determined by ecosystems. We therefore perceive the greatest need for research to lie in some of the areas discussed below.

SUSTAINING DIVERSITY VERSUS ISOLATING DISTURBANCES

Complex systems are constantly poised between the opposing forces of diversification and selection. In ecosystems such questions are often posed in terms of species present and species lost and the influence of species loss on ecosystem function. In social systems the challenge becomes one of sustaining a diversity of the mental models that people use to map different actions to their predicted outcomes while at the same time enabling the sharing of experiences and the promotion of innovation through the processes of interacting with new people who have different perspectives. Many of the important dynamics of social-ecological systems, such as mobilizing innovation and knowledge when needed, do not come from static structures, but come instead from the creation and removal of links between diverse nodes and the learning and the updating of mental models that take place within nodes. This perspective leads to a number of hypotheses. For example, finding new solutions by participatory approaches is not achieved by having denser social networks; over time this would be expected to lead either to homogenization of mental models or to polarization and conflict, depending on the social processes and the situation (Scheffer, Westley, and Brock 2003). Static high connectivity in networks is expected to lead to erosion of diversity in mental models as experiences are shared among all participants (Bodin and Norberg 2005), while low static connectivity is expected to lead to inefficient use of knowledge. Some of the central research questions in this context include understanding how different kinds of social-ecological network dynamics influence the properties of the network, testing alternative ways to both sustain and mobilize diversity, and understanding the ways in which human social interactions influence adaptation, innovation, and the long-term resilience of the social-ecological system.

SOCIAL ATTRACTORS AND REGIME SHIFTS

The concept of regime shifts as a fundamental change in system-internal feedbacks has gained considerable interest in ecology, but much less so in the social sciences (Carpenter and Brock 2004). One reason for this is semantic, i.e., regimes in the social sciences cover the concept of a governance structure rather than a description of alternate trajectories of development. We suggest that regimes and regime shifts could be useful conceptual tools for understanding the social aspects of natural resource management and are in need of further research. In the natural sciences a regime is identified by a particular feedback process that will drive a system toward a (potentially dynamic) end-state. A regime in the social context could thus be defined as decision-outcome processes in which the choices for particular solutions are reinforced by their outcomes, i.e., the effects on the preferences of the people that influence the decision-making process. For example, if a wetland is considered a mudhole with little value, decisions at the municipal or individual level will not consider its preservation a top priority, leading to actions that most likely will decrease its value even further. By contrast, if the wetland is considered to be of high value to local people because of recreational or aesthetic considerations, decisions at municipal and individual levels will likely prioritize its preservation and management and thus potentially increase its value even further. These alternatives represent two contrasting and self-reinforcing regimes. A real-life example of how this thinking has been applied in wetland management can be found in (Olsson, Folke et al. 2004b) and in chapter 4 of this volume. In southern Sweden a small number of key individuals were able to lead social changes that transformed a regime in which the wetland was considered a water-sick area to a regime in which the same wetland area is now internationally marketed as a water kingdom or eco-museum.

Research on the role of preferences will be critical to understanding social regimes and their interactions with ecosystems. The preferences of people influence decisions and are crucial for determining the outcomes of individual and collective choice. As discussed in chapter 5, however, preferences are sometimes only apparent (i.e., conceived without any deliberation) and may even be counterproductive for deciding the best action to take in increasing an individual's well-being. In such cases stewards or institutional entrepreneurs (Westley and Miller 2003) may provide simple visions or scenarios to help people understand more complex relationships and to contribute to more long-term and sustainable decisions (chapter 8, Lebel and Bennett). Visioning and scenario building are methods that help to describe and shape attractors in social-ecological systems, but their overall contribution to social-ecological dynamics has not been researched and there have been few attempts to document their successes and failures.

GOVERNING SELF-ORGANIZATION

Democratic winner-takes-all processes have limitations for the governance of SES, but few viable alternatives exist. While bottom-up approaches are commendable, and participation and collective action may utilize local knowledge effectively, these approaches to governance are also sensitive to the personalities and objectives of local leaders. Self-organizing social processes can be dominated by a single person who can affect the preferences of others and monopolize visions. Some of the central research questions in this context include understanding how social and ecological self-organization occurs, how social-ecological dynamics differ from social or ecological dynamics, and the different ways in which social-ecological self-organization can be managed. Governing self organization involves an understanding of the diversity of components and their interactions. Network approaches may provide good tools for mapping out important interactions and critical components, but a static picture of interactions does not tell much about the dynamics of the system when faced with a crisis or continual change. To understand this, the responses of the nodes, that is, the information processing capabilities, need to be understood, at least for the most critical nodes.

SPATIAL VARIATION IN SOCIAL-ECOLOGICAL DYNAMICS

Both social systems and ecological systems are composed of patterns and processes that occur and interact across multiple scales. Each kind of system varies through space, with regional and local differences in the structure and function of ecosystems and the composition, institutions, and governance of social systems. These spatial differences lead to varied social-ecological interactions in different locations and can have a large effect on such properties of a system as its ability to adapt to change, the location of thresholds between different regimes, and the processes that dominate regime shifts. For example, farmers in drought-prone regions tend to have fewer options for crop farming and must generally deal with greater temporal variability in their food supply, meaning that they must maintain a base of capital to see them through bad years, consequently reducing their range of options for other kinds of investment. At a practical level spatial differences influence the generality of models of SESs and the exportability of locally developed solutions to other contexts. As highlighted in chapter 1 (Cumming, Barnes, and Southworth), further research is needed to understand the relevance of location and connectivity between SESs, the ways in which the local and regional environment constrains or drives SES dynamics, and the ways in which spatial variation within and between different systems influences system resilience. The topic of spatial variation is also relevant to understanding the relationships between processes at different scales of social-ecological systems and the possibilities for resolving scale mismatches, as discussed in detail in chapter 9 (Cumming and Norberg).

The challenge we face in securing development toward a sustainable future is to comprehend the integrated nature of the world we live in, the feedbacks and responses within the social-ecological system and the multilevel cause and effect relationships that govern its dynamics. Change and disturbance play a central role in our ability to learn and sustain the capacity to respond to surprise; attempts at controlling change and disturbance will backfire as this capacity is eroded. Disturbance and crisis are integral to development toward better solutions. A central focus needs to be put on the processes that affect the diversity within a system, since these affect our future options. This includes biodiversity as well as the diversity of local ecological knowledge, institutional diversity, and cultural diversity.

In order to enable solutions instead of providing them in a top-down fashion, new tools of management need to be established. These should include 1. altering the network of interaction, e.g., by facilitating bridging and bonding of nodes if a denser network is desired, 2. sustaining the processes that maintain the diversity of components in the system, such as the variety of species, livelihoods, or mental models of people, for example, by facilitating exchange between regions, and 3. identifying the the nodes that are more important than others and providing targeted management. The later could for example include stepping-stone patches in a network of nature reserves or providing a forum for ecological entrepreneurs that have gained trust in the community and can help build a common sustainable vision together with policy makers. Identifying the sources of diversity—how they interact across different scales and how they feed back into the processes that determine the trajectory of development of social-ecological systems—presents a massive challenge for sustainability research in the coming years.

This volume illustrates the ways in which complexity theory is now reaching a stage where it is of high relevance to societies and ecosystems. Although considerable progress has been made in developing and exploring different applications of complexity theory to the management of social-ecological systems, much remains unknown and there is an enormous potential for innovative and exciting research. It is our hope that this book will offer a worthwhile contribution to the gradual transformation of conservation biology, as well as related areas of sociology and economics, and that it will help to foster a broader appreciation of the importance of interdisciplinary interactions for the further development of the theory and management approaches that are needed to guide our society into a sustainable future.

REFERENCES

Arrow, H., J. E. McGrath, and J. L. Berdahl. 2000. *Small Groups as Complex Systems: Formation, Coordination, Development, and Adaptation.* Thousand Oaks, CA: Sage.

Berkes, F., J. Colding, and C. Folke. 2003. *Navigating Social-Ecological Systems: Building Resilience for Complexity and Change.* Cambridge: Cambridge University Press.

Berkes, F., and C. Folke. 1998. *Linking Social and Ecological Systems: Management Practices and Social Mechanisms for Building Resilience.* Cambridge: Cambridge University Press.

Bodin, Ö., and J. Norberg. 2005. "Information Network Topologies for Enhanced Local Adaptive Management." *Environmental Management* 35(2): 175–193.

Briske, D. D., S. D. Fuhlendorf, and F. E. Smeins. 2003. "Vegetation Dynamics on Rangelands: A Critique of the Current Paradigms." *Journal of Applied Ecology* 40(4): 601–614.

Brock, W., and S. N. Durlauf. 2005. "Local Robustness Analysis: Theory and Applicaton " *Journal of Economic Dynamics and Control* 29:2067–2092.

Brock, W., and A. Xepapdeas. 2004. "Management of Interacting Species: Regulation Under Nonlinearities and Hysteresis." *Resource and Energy Economics* 26:137–156.

Carlson, J. M., and J. Doyle. 1999. "Highly Optimized Tolerance: A Mechanism for Power Laws in Designed Systems." *Physical Review E* 60:1412–1427.

Carpenter, S. R. 2003. *Regime Shifts in Lake Ecosystems: Pattern and Variation.* Oldendorf/Luhe: International Ecology Institute.

Carpenter, S. R., and W. A. Brock. 2004. "Spatial Complexity, Resilience, and Policy Diversity: Fishing on Lake-Rich Landscapes." *Ecology and Society* 9(1): 8.

Carpenter, S. R., and W. A. Brock. 2006. "Rising Variance: A Leading Indicator of Ecological Transition." *Ecology Letters* 9:311–318.

Carpenter, S. R., and L. H. Gunderson. 2001. "Coping with Collapse: Ecological and Social Dynamics in Ecosystem Management." *BioScience* 6:451–457.

Clark, W. C., and N. M. Dickson. 2003. "Sustainability Science: The Emerging Research Program." *Proceedings of the National Academy of Sciences of the United States of America* 100(14): 8059–8061.

Costanza, R. 1991. *Ecological Economics: The Science and Management of Sustainability.* New York: Columbia University Press.

Dasgupta, P. 2001. *Human Well-Being and the Natural Environment.* Oxford: Oxford University Press.

Davis, M. B. and R. G. Shaw. 2001. "Range Shifts and Adaptive Responses to Quaternary Climate Change." *Science* 292(5517): 673–679.

Diamond, J. 2005. *Collapse: How Societies Choose to Fail or Survive.* London: Allen Lane.

Dietz, T., E. Ostrom, and P. C. Stern. 2003. "The Struggle to Govern the Commons." *Science* 302:1902–1912.

Ehrlich, P. R. and A. H. Ehrlich. 1981. *Extinction: The Causes and Consequences of the Disappearance of Species.* New York: Random House.

Elmqvist, T., C. Folke, M. Nyström, G. Peterson, J. Bengtsson, B. Walker, and J. Norberg. 2003. "Response Diversity and Ecosystem Resilience." *Frontiers in Ecology and the Environment* 1:488–494.

Folke, C. 2006. "Resilience: The Emergence of a Perspective for Social-Ecological Systems Analyses." *Global Environmental Change* 16:253–267.

Folke, C., T. Hahn, P. Olsson, and J. Norberg. 2005. "Adaptive Governance of Social-Ecological Systems." *Annual Review Environment and Resources* 30:441–473.

Goldman, A. 1995. "Threats to Sustainability in African Agriculture: Searching for Appropriate Paradigms." *Human Ecology* 23(3): 291–334.

Holling, C. S. 1992. "Cross-Scale Morphology, Geometry, and Dynamics of Ecosystems." *Ecological Monographs* 62(4): 447–502.

Holling, C. S. and G. K. Meffe. 1996. "Command and Control and the Pathology of Natural Resource Management." *Conservation Biology* 10(328–337).

Jackson, J. B.C., M. X. Kirby, W. H. Berger, K. A. Bjorndal, L. W. Botsford, B. J. Bourque, R. H. Bradbury, R. Cooke, J. Erlandson, J. A. Estes, T. P. Hughes, S. Kidwell, Carina B. Lange, H. S. Lenihan, J. M. Pandolfi, C. H. Peterson, R. S. Steneck, M. J. Tegner, Robert R. Warner. 2001. "Historical Overfishing and the Recent Collapse of Coastal Ecosystems." *Science* 293(5530): 629–637.

Kates, R. W., and W. C. Clark. 1996. "Expecting the unexpected." *Environment* 38:6–18.

Levin, S. A. 2002. "Complex Adaptive Systems: Exploring the Known, the Unknown and the Unknowable." *Bulletin New Series of the American Mathematical Society* 40(1): 3–19.

Ludwig, D., R. Hilborn, C. Walters. 1993. "Uncertainty, Resource Exploitation, and Conservation: Lessons from History." *Science* 260(2): 17.

MEA. 2005. *Biodiversity and Human Well-being*. Millennium Ecosystem Assessment. http://www.millenniumassessment.org/.

Misztal, B. A. 1996. *Trust in Modern Societies*. Cambridge: Polity.

Olsson, P., C. Folke, and T. Hahn. 2004b. "Social-Ecological Transformation for Ecosystem Management: The Development of Adaptive Co-management of a Wetland Landscape in Southern Sweden." *Ecology and Society* 9(4): 2.

Ostrom, E. 2005. *Understanding Institutional Diversity*. Princeton: Princeton University Press.

Pauly, D. 1995. "Anecdotes and the Shifting Baseline Syndrome of Fisheries." *Trends in Ecology and Evolution* 10(10): 430.

Putnam, R. D. 2004. *Democracies in Flux: The Evolution of Social Capital in Contemporary Society*. Oxford: Oxford University Press.

Robert, C., J. M. Carlson, and J. Doyle. 2001. "Highly Optimized Tolerance in Models Incorporating Local Optimization and Regrowth." *Physical Review E* 63:56122–56135.

Scheffer, M., F. Westley, and W. Brock. 2003. "Slow Response of Societies to New Problems: Causes and Costs." *Ecosystems* 6:493–502.

Skogestad, S., and I. Postlethwaite. 1996. *Multivariable Feedback Control: Analysis and Design*. New York: Wiley.

Vitousek, P. M., H. A. Mooney, J. Lubchenko, and J. M. Melillo. 1997. "Human Domination of Earth's Ecosystems." *Science* 277:494–499.

Walker, B., and J. Meyers. 2004. "Thresholds in Ecological and Social-Ecological Systems: A Developing Database." *Ecology and Society* 9(2): 3.

Westley, F. 2002. "The Devil in the Dynamics: Adaptive Management on the Front Lines." In L. Gunderson and C. S. Holling, eds., *Panarchy: Understanding Transformations in Human and Natural Systems*, pp. 333–360. Washington, DC: Island.

Westley, F., and P. S. Miller. 2003. *Experience in Consilience: Integrating Social and Scientific Responses to Save Endangered Species*. Washington, DC: Island.

Westley, F., S. R. Carpenter, W. A. Brock, C. S. Holling, and L. H. Gunderson 2002. "Why Systems of People and Nature Are Not Just Social and Ecological Systems." In L. H. Gunderson and C. S. Holling, eds., *Panarchy: Understanding Transformations in Human and Natural Systems*, pp. 103–119. Washington, DC: Island.

INDEX

Printed in the USA
CPSIA information can be obtained
at www.ICGtesting.com
JSHW021435221024
72172JS00002B/11

9 780231 134613